AF875989

Mauricio Lobos Fernández

Homogenization and materials design of mechanical properties of textured materials based on zeroth-, first- and second-order bounds of linear behavior

Schriftenreihe
Kontinuumsmechanik im Maschinenbau
Band 12

Karlsruher Institut für Technologie (KIT)
Institut für Technische Mechanik
Bereich Kontinuumsmechanik

Hrsg. Prof. Dr.-Ing. habil. Thomas Böhlke

Eine Übersicht aller bisher in dieser Schriftenreihe erschienenen Bände finden Sie am Ende des Buchs.

Homogenization and materials design of mechanical properties of textured materials based on zeroth-, first- and second-order bounds of linear behavior

by
Mauricio Lobos Fernández

KIT Scientific Publishing

Dissertation, Karlsruher Institut für Technologie
KIT-Fakultät für Maschinenbau

Tag der mündlichen Prüfung: 29. Januar 2018
Referenten: Prof. Dr.-Ing. habil. Thomas Böhlke
Prof. Dr. habil. Adam Morawiec
Prof. Dr. habil. Helmut Schaeben

Impressum

KIT Scientific Publishing

Karlsruher Institut für Technologie (KIT)
KIT Scientific Publishing
Straße am Forum 2
D-76131 Karlsruhe

KIT Scientific Publishing is a registered trademark
of Karlsruhe Institute of Technology.
Reprint using the book cover is not allowed.

www.ksp.kit.edu

This document – excluding the cover, pictures and graphs – is licensed under a Creative Commons Attribution-Share Alike 4.0 International License (CC BY-SA 4.0): https://creativecommons.org/licenses/by-sa/4.0/deed.en

The cover page is licensed under a Creative Commons Attribution-No Derivatives 4.0 International License (CC BY-ND 4.0): https://creativecommons.org/licenses/by-nd/4.0/deed.en

Print on Demand 2018 – Gedruckt auf FSC-zertifiziertem Papier

ISSN 2192-693X
ISBN 978-3-7315-0770-3
DOI 10.5445/KSP/1000080683

Homogenization and materials design of mechanical properties of textured materials based on zeroth-, first- and second-order bounds of linear behavior

Zur Erlangung des akademischen Grades

Doktor der Ingenieurwissenschaften

der Fakultät für Maschinenbau

Karlsruher Institut für Technologie (KIT)

genehmigte

Dissertation

von

M.Sc. Mauricio Lobos Fernández

Tag der mündlichen Prüfung:	29.01.2018
Hauptreferent:	Prof. Dr.-Ing. habil. Thomas Böhlke
Korreferenten:	Prof. Dr. habil. Adam Morawiec
	Prof. Dr. habil. Helmut Schaeben

Zusammenfassung

Die vorliegende Arbeit setzt sich (1) mit der Homogenisierung linearer und nichtlinearer mechanischer Eigenschaften von Materialien mit kristallographischer und morphologischer Textur, sowie (2) mit dem Materialdesign mechanischer Eigenschaften bei vorgeschriebenem Eigenschaftenprofil auseinander. Für die grundlegende Beschreibung der Textur im Material wird die Orientierungsverteilungsfunktion von Kristalliten verwendet. Diese grundlegende Größe der Mikrostruktur texturierter Materialien erscheint auf natürliche Art und Weise in zahlreichen Homogenisierungsansätzen für verschiedene anwendungsrelevante Materialien (z.B. partikelverstärkte Komposite und Polykristalle mit anisotropen Materialverhalten). Aufgrund der Vielfältigkeit und Komplexität von Texturen, ist es für die Anwendung notwendig, möglichst effiziente und niedrig dimensionale Darstellungen von texturabhängigen Homogenisierunganŝätzen zu entwickeln. Dies wird in der vorliegenden Arbeit basierend auf Darstellungen in Abhängigkeit von *tensorwertigen Texturkoeffizienten* erreicht. Die Darstellungen dieser Arbeit liefern für die linear elastischen Eigenschaften beliebig anisotroper Materialien Schranken und Approximationen nur in Abhängigkeit von Texturkoeffizienten *zweiter und vierter Ordnung*, welche eine *endlich dimensionale Parametrisierung in konvexen Gebieten* ergeben. Dazu werden die in der Literatur bekannten Schranken linearer Eigenschaften von Voigt, Reuss (Schranken erster Ordnung) und Hashin-Shtrikman (Schranken zweiter Ordnung) unter Berücksichtigung von Eigenfeldern betrachtet. Zusätzlich zu der texturabhängigen Parametrisierung dieser Schranken, werden die Schranken nullter Ordnung mit Berücksichtigung von Eigenfeldern abgeleitet. Die Schranken nullter Ordnung sind materialabhängig, aber mikrostrukturunabhängig. All diese Ergebnisse werden in dieser Arbeit im Zusammenhang mit einem Materialdesign linearer und nichtlinearer Eigenschaften anhand von drei Beispielen präsentiert. Im ersten Beispiel werden die elastischen Eigenschaften eines orthotropen Polykristals eines kubischen Materials diskutiert. Hierbei werden, basierend auf den Schranken nullter Ordnung und Datenbanken, passende Materialkandidaten entsprechend dem geforderten Eigenschaftenprofil ausgewählt. Dann werden die Schranken erster und zweiter Ordnung ausgewertet, sowie eine darauf aufbauende Approximation in Anhängigkeit der Texturkoeffizienten. Vorteilhafte Texturkoeffizienten werden basierend auf der Approximation ermittelt. Im zweiten Beispiel werden (analog zum ersten) die linear thermoelastischen Eigenschaften eines transversalisotropen partikelverstärkten Komposits ausgewertet. Hierbei wird die Anwendung der texturabhängigen Ausdrücke mit Berücksichtigung von Eigenfeldern demonstriert. Darüber hinaus wird diskutiert, wie eine morphologische Textur durch Modellannahmen als kristallographische Textur modelliert werden kann. Im letzten Beispiel wird die Anwendung der texturabhängigen Hashin-Shtrikman Schranken im Kontext der Schranken vom nichtlinearen viskoplastischen Materialverhalten von Polykristallen demonstriert. Hierfür werden mithilfe von Variationsprinzipien (basierend auf linearen Vergleichsmaterialien) texturabhängige untere Schranken für das effektive Fließpotential und obere Schranken für die effektive Fließspannung abgeleitet. Ein analytischer Sonderfall für Polykristalle von kubisch-flächenzentrierten Materialien wird angegeben und diskutiert.

Summary

The present work approaches (1) the field of homogenization of linear and nonlinear mechanical properties of materials with crystallographic and morphological texture, as well (2) the field of materials design for the mechanical properties with prescribed properties-profile. The basic material texture is described by the crystallite orientation distribution function. This essential microscopic quantity of textured materials appears naturally in a large number of homogenization approaches of different application relevant materials, e.g., particle reinforced composites and polycrystals with anisotropic material behavior. Due to the wealth and complexity of textures, it is essential for applications to develop efficient and low dimensional representations of texture dependent homogenization approaches. This is achieved in this work based on representations in terms of *tensor valued texture coefficients*. For arbitrarily anisotropic linear elastic material behavior, the representations deliver bounds and approximations in terms of solely *second- and fourth-order texture coefficients*, which yield *finite-dimensional parametrizations in convex sets*. The well-known bounds of Voigt, Reuss (first-order bounds), and Hashin-Shtrikman (second-order bounds) are taken into consideration, accounting for eigenfields. In addition to the texture dependent parametrizations of these bounds, the zeroth-order bounds accounting for eigenfields are derived. These bounds of the linear material behavior are material dependent and microstructure independent. All these results are then presented in the context of the materials design of linear and nonlinear properties in three examples. In the first example, the linear elastic properties of an orthotropic polycrystal of a cubic material are discussed.

For this purpose, based on the zeroth-order bounds, material candidates are selected from a database according to the prescribed properties-profile. Then, the first- and second-order bounds are evaluated for the delineation of the properties-closures, together with a synthesizing approximation in terms of texture coefficients. Favorable texture coefficients are determined based on the approximation. In the second example, following the first example, the linear thermoelastic properties of a transversely isotropic particle reinforced composite are evaluated. Here, the application of the texture dependent expressions accounting for eigenfields of this work is demonstrated, as well as modeling for the treatment of morphological texture as crystallographic texture. In the last example, the application of the texture dependent Hashin-Shtrikman bounds in the context of bounds of nonlinear viscoplastic material behavior of polycrystals is demonstrated. Based on variational principles (using a linear comparison material) texture dependent lower bounds of the effective flow potential are derived, together with texture dependent upper bounds for the effective flow stress. One analytic special case for FCC polycrystals is presented and discussed.

Acknowledgments

I want to first thank Prof. Thomas Böhlke for bringing micro-mechanics and analytic homogenization to my attention since my Bachelor's studies. Prof. Böhlke always kept me interested in these subjects, offered me guidance and freedom for my research, for which I am sincerely grateful.

The Karlsruhe House of Young Scientists (KHYS) is gratefully acknowledged for the funding of my research stay at the Department of Applied Mathematics and Theoretical Physics (DAMTP) in Cambridge, England, visiting Prof. John R. Willis.

Furthermore, I would like to express my gratitude to Prof. Adam Morawiec and Prof. Helmut Schaeben for co-advising and improving this work with their valuable suggestions.

I also want to thank all my great work colleagues at the ITM, the old brigade and the next generation, for the support, the discussions, the extensive meals, the coffee and the wonderful friendships (with all the stories from conferences, vacations and city tours).

Finally, I want to thank mi amor Hanna for accepting and appreciating me as the nerd that I am and my beloved family in Germany and Chile, who has supported and encouraged me not only during my thesis, but throughout every aspect of my life. Especially, I want to thank my mother Blanca Rosa Fernández Peirano for always being there for me, for raising me with love, compassion and understanding. I would not be what I am today, if I would not have had her.

Karlsruhe, May 2018 Mauricio Lobos Fernández

Contents

Chapter 1

Introduction

1.1 Motivation

Modern materials research is concerned with the manufacturing of multiphase materials, i.e., materials containing, in general, several material constituents (referred to as phases) and a heterogeneous spatial distribution of the material constituents (referred to as microstructure). For example, for many automotive applications, metals are heavily used in several components of cars. The material behavior of the metal is governed by the material and its microstructural arrangement. At a microscopic level, one can see that most metals might be described as an aggregate of single crystals, i.e., a polycrystal. The anisotropic material is the same in each crystallite, but its orientation differs from crystallite to crystallite. This material orientation distribution of the crystallographic axes of the reference material is referred to as crystallographic texture. Although a metal of a pure material possesses only a single material constituent, the material stiffness varies in space from crystallite to crystallite due to the different material orientations in the crystallites. Each distinct material orientation connected to the corresponding crystallites in the polycrystal is to be considered as a distinct phase in the polycrystal. Therefore, a polycrystal of a single material might be referred to as a multiphase material with a large or even infinite number of phases. In this work, we will instead refer to such a material as a single-phase polycrystalline material (SPPM) to emphasize the material constituent and polycrystalline nature of the polycrystal. Other examples of heterogeneous materials are fiber reinforced lightweight materials and matrix-particle mixtures. These materials might be composed of solely isotropic constituents, but the spatial arrangement of the fibers or particles having specific geometry in the embedding matrix might be highly complex. This is referred to as morphological texture since the orientation distribution of the shapes is considered. If the phases of a multiphase material are anisotropic and polycrystalline, then we refer to the material as a multiphase polycrystalline material (MPPM) which may show crystallographic and morphological texture. In this work, we will shortly refer to all of these materials as textured materials.

The space of properties offered by different material combinations for different physical problems can be very rich if either anisotropic materials or material combinations with high material contrast (e.g., a weak matrix with stiff inclusions) are considered. Purposefully chosen microstructures can induce highly preferable properties for different applications. But, inventing different microstructures and testing materials is a procedure with too many costs regarding resources and time. Therefore, theoretical investigations of the behavior of materials are necessary in order to accelerate the discovery of materials for specific applications, where different physical prescribed material properties with imposed tolerances, referred to as properties-profile throughout this work, are of interest. With this goal in mind, the theory of multiphase heterogeneous materials needs to answer at least three central questions:

- Q1 What are the effective properties of a material with a given microstructure and given local material properties?
- Q2 Is it possible for a chosen/given material with variable microstructure to reach a required properties-profile?
- Q3 Which microstructures of a given material are favorable for the fulfillment of the desired properties-profile?

The first question (Q1) deals with the problem of homogenization. It follows the procedure of considering given microstructures and computation of the effective material properties of heterogeneous materials, usually based on as little statistical microstructure-information as possible. The homogenization of materials with complex microstructures is a challenging problem which usually requires a high amount of statistical microstructure-information to deliver accurate results. Therefore, a statistical description of microstructures is essential in any case concerning such approximations. How much statistical information is needed depends on the properties of interest for the application and the material contrast of the chosen materials. Material combinations with high material contrast show significant differences in their directional properties if the microstructure is changed.

Materials design is referred to in this work as the optimization of material and microstructure depending on a prescribed properties-profile. Material designers might have to confront other problems besides homogenization. Homogenization has the disadvantage of computing a number, e.g., 10, and then asking the world, if anybody needs a 10. Material designers rather work *inversely*, they ask themselves: "I need a 42. What gives me 42? What is *even able* to give me 42?". Therefore, the first question, a material designer should ask himself is, if a material can deliver properties required for a specific application. This is the problem addressed by the question Q2. A chosen material, e.g., glass, might offer high stiffness values. But for the properties-profile needed for a precise application these values might be, perhaps, inadequate or too high. Therefore, glass is *not able* to deliver properties in the range required for the application of interest. This raises the alternative question: has a material intrinsic lower and upper limitations? This question can be approached in theory for certain materials, such that lower and upper bounds of potentials can be derived based on physical principles, which induce lower and upper bounds for the material properties for the physical problem at hand. Such lower and upper bounds do not give explicit information about the exact material properties. Bounds give information about *definite limitations of the material*, and *what is definitely not reachable by the material*. This is crucial for material designers. For many cases, it cannot be shown if the chosen material is able to reach specific properties, but it can be shown that the selected material is *not* able to achieve the properties-profile. This *excluding* property of bounds allows material designers to exclude materials and microstructures based on required properties-profile, such that inappropriate materials in the databases can be immediately discarded. This would save high amounts of otherwise unnecessarily wasted resources in manufacturing, testing and/or simulating materials which would never be able to reach required properties.

The last question (Q3) is the hardest of the three questions in the purposeful engineering of materials. Two approaches can be considered. In the first one, a sensible approximation of the effective material is required which does not violate any bounds derived from the physical problem at hand. Based on a pragmatic point of view, such approximations usually fulfill bounds only up to a certain order, since otherwise the exact material behavior would be needed, which is normally not known. Further, these approximations should depend on as little statistical microstructure-information as possible and given in *finite-dimensional* parametrizations. If an approximation has these properties, then it might be able to deliver practical results in optimization of its microstructural variables concerning the prescribed properties-profile. Based on the optimized microstructural variables, the bounds considered might be evaluated to check how far deviations might occur. This first approach is the simplest possible. In the second approach, considered bounds can be used in order to scan the space of microstructural variables for sets inducing properties bounds with non-empty intersections with the properties-profile. This problem is the most complex one since the evaluation of higher-order bounds might reveal that no microstructures exist which deliver properties bounds intersecting the properties-profile. These three questions motivate the investigations of the present work for polycrystalline multiphase materials with linear and nonlinear mechanical material behavior.

1.2 State of the art

The theory of the effective behavior of heterogeneous materials has a vast literature. The linear elastic behavior of, e.g., polycrystals has been investigated already by Voigt (1910) and Reuss (1929), which proposed to evaluate the volume averages of the stiffness (assuming constant deformation) and compliance (assuming constant stresses), respectively. These approaches have been shown by Hill (1952) to deliver actual bounds of the linear elastic material behavior depending on volume fraction information of the heterogeneous materials. Based on two-point statistical information of the microstructure, second-order bounds of the linear elastic properties can be formulated, e.g., based on the Hashin-Shtrikman variational principle of Hashin and Shtrikman (1962), see also Walpole (1966) and Willis (1977). If higher-order statistical information is taken into account, then higher-order bounds and approximations based on these bounds can be computed, see, e.g., Kröner (1977) and Willis (1981). One of the more frequently used approximations (based on bound expressions and derivation) in linear elasticity is the so-called self-consistent approach, see Kröner (1958), Kröner (1977), Willis (1977) and Willis (1981).

For different applications, physical problems with simplified coupled behavior are also of interest. One example is linear thermoelasticity, where the absolute temperature is assumed to be constant throughout the material, and the mechanical material response of the composite is to be examined. These kinds of problems have been analyzed, e.g., by Rosen and Hashin (1970) and Laws (1973), where bounds and self-consistent approximations are derived. This problem can be viewed, more generally, as a linear elastic material with eigenfields (eigenstresses or eigenstrains), see, e.g., Willis (1981). Bounds of this problem type are known explicitly only for simplified, usually isotropic cases, due to the cumbersome algebraic expressions for anisotropic materials or even materials with orientation dependency, e.g., textured materials.

Based on the homogenization and bounding theory of linear material behavior, different models and bounds for nonlinear behavior have been developed. In Taylor (1938) the deformation is assumed constant (Voigt assumption) in polycrystals with nonlinear material response, which deliver an upper bound of the effective elastic potential, see, e.g., Bishop and Hill (1951) or Dendievel et al. (1991). The dual assumption of constant stresses (Reuss assumption) delivers an upper bound of the effective complementary potential, see, e.g., Dendievel et al. (1991). Bounds of nonlinear material behavior based on the Hashin-Shtrikman variational principle have been derived following Willis (1983) by Talbot and Willis (1985). In Ponte Castañeda (1991) an even more general approach for bounding nonlinear properties has been introduced. This approach is highly attractive for a wide part of the community of homogenization since it is based on comparing the potential of the nonlinear material with a potential corresponding to a linear material, referred to as linear comparison composite. This gives access to the complete theory of linear materials for bounding nonlinear behavior. The approach of Ponte Castañeda (1991) can be extended to consider eigenfields in the linear comparison composite, see, e.g., Ponte Castañeda et al. (2004), such that auxiliary bound expressions for such materials, like the ones presented in Willis (1981), can be used in nonlinear context. This approach continues to be extended and optimized, see, e.g., Ponte Castañeda (2015) and Ponte Castañeda (2016), but with no efficient *parametrization* for materials showing complex directional dependency, as, e.g., textured materials.

All of the just addressed bounds and approximations for linear and nonlinear material behavior are adequate (regarding computational time) for calculations involving rather isotropic behavior. Especially, bounds and approximation of textured polycrystals, having an intrinsic directional dependency due to the underlying anisotropic single crystal behavior and its orientation distribution over the spatial domain of interest, are challenging to evaluate. The orientation distribution of the single crystal behavior is the most basic statistical description of the microstructure of polycrystals and other textured materials. But, bounds and approximations of the material response depending on it are already algebraically difficult to evaluate, if represented by a simple convex combination of single crystal states. In the community of quantitative texture analysis, the crystallite orientation distribution function (CODF) is a well studied quantity, for which advantageous representations based on Fourier series in terms of so-called texture coefficients are known, see, e.g., Bunge (1982), Adams et al. (1992), Guidi et al. (1992), Schaeben and van den Boogaart (2003) and Adams et al. (2013). Alternatively, CODFs can be approximated by simplified model functions, see, e.g., Matthies et al. (1988), Schaeben (1992) and Schaeben (1996), which simplify the texture coefficients drastically. These representations give access to highly advantageous parametrizations of, e.g., the first-order bounds of linear elastic properties (see, e.g., Adams et al. (2013)) due to the dependency on texture coefficients. These coefficients enter several elasticity relevant expressions in a finite-dimensional fashion. This property is a crucial simplification which opens the fast and efficient exploration of properties delivered by bounds and approximations parametrized with texture coefficients. Naturally, corresponding bounds (and approximations) depending on the CODF do not need to be expressed in terms of texture coefficients. All corresponding bounds can be parametrized by the CODF, but this parametrization has the disadvantage of taking into consideration the infinite-dimensional function space of all CODFs. This makes the exploration of the respective bounds severely more complicated, compared to equivalent representations regarding finite-dimensional texture coefficients.

Modern problems in materials design, as indicated in Section 1.1, are *inverse* in nature. The objective is not to compute the effective properties of a given material and a given microstructure, but to optimize material and microstructure variables according to a given properties-profile. First, in the selection of a material, large databases might be available to material designers, see, e.g., de Jong et al. (2015). The selection of suitable anisotropic materials or their combinations is already a huge problem. If no physical and fast computable bounds for the anisotropic case are at hand, then poor materials selection based only on isotropic cases is the only pragmatic option in large material databases. And this is the case which only considers changing the underlying material constituents, i.e., the variations depending on microstructure adaption has not been even taken into account. If the material is somehow given/chosen, then bounds and approximations given in terms of *finite-dimensional* representations might be adequate for efficient computations, as needed in inverse problems. Approaches based on texture coefficients of the CODF are presented in, e.g., Adams et al. (2001), Kalidindi et al. (2004), Proust and Kalidindi (2006), Fullwood et al. (2010) and Adams et al. (2013) for linear and nonlinear properties based on the Voigt and Reuss bounds, approximations and Fourier series of the material response. The author is not aware of the incorporation of second-order bounds for linear material behavior and corresponding bounds for eigenfields (as the Hashin-Shtrikman bounds of Willis (1977) and Willis (1981)) or nonlinear behavior (based on approaches as the one introduced in Ponte Castañeda (1991)) in the framework of materials design accounting for textured materials, see, e.g., Fullwood et al. (2010) or Adams et al. (2013).

1.3 Objectives and outline of this work

Based on the questions discussed in Section 1.1 and the requirements of materials design problems, the main objectives of this work are:

1. The investigation of material dependent but microstructure independent bounds with a fast evaluation for large material databases.
2. The derivation of efficient expressions for the second-order Hashin-Shtrikman bounds accounting for eigenfields in terms of texture coefficients for multiphase polycrystalline materials of arbitrary anisotropy.
3. The application of texture dependent Hashin-Shtrikman expressions for linear behavior in the context of nonlinear approaches.

These objectives require a compact description of the concept of orientations, their parametrizations, distribution and relevant influence on the orientation average of physical tensorial properties of solid materials, which is provided in Chapter 2. This a priori preparation will drastically simplify the readability and the treatment of homogenization of linear and nonlinear problems of textured materials discussed in the subsequent chapters.

The basic homogenization theory needed for the present work is sketched in Chapter 3. The chapter starts with a short description of the microstructure of heterogeneous materials, followed by a brief illustration of basic homogenization of random materials. Subsequently, some general relations of the theory of bounds for linear problems are discussed, together with zeroth-, first- and second-order bounds (all accounting for eigenfields). The zeroth-order bounds are presented to some length to show their potential application in large material databases in material design problems. The first-order bounds are recapitulated rather shortly since these are well known in literature. A more extensive treatment is given to the second-order Hashin-Shtrikman bounds for isotropic two-point statistics to illustrate their derivation and explicit texture dependency. At the end of the chapter, the potential application of all derived expressions is discussed for linear thermoelasticity and the approach introduced by Ponte Castañeda (1991) but accounting for arbitrary texture.

In order to show possible application scenarios within the field of materials design, the results of the present work are illustrated in Chapter 4 in three examples. In the first example, a polycrystal of a single cubic material showing orthotropic texture will be designed based on a prescribed properties-profile and given material database. In the second example, the linear thermoelastic properties of a matrix-inclusion composite with a weak isotropic matrix and transversely isotropic inclusions with transversely isotropic macroscopic behavior will be optimized based on a prescribed properties-profile and material database. Finally, in the last example, the nonlinear behavior of a viscoplastic polycrystal with power-law behavior will be bounded with the texture dependent expressions of the present work. The special case of an FCC material with cubic macroscopic texture is illustrated to present results computable by hand. For this case, compact bounds for the resulting nonlinear potential and effective flow stress are derived and discussed.

At the beginning and end of each chapter, an overview and a review of the chapter is provided in order to improve readability and keep the main objectives and results as clear as possible throughout this work.

1.4 Notation

In the present work, a direct tensor notation is preferred throughout the text. Einstein summation convention is *not* used in this work. Scalars are denoted by standard italic characters, e.g., x, y, W. Lower case bold Latin characters, e.g., $\boldsymbol{x}, \boldsymbol{y}$, denote first-order tensors, while upper case bold Latin characters, e.g., $\boldsymbol{A}, \boldsymbol{B}$, and bold Greek characters, e.g., $\boldsymbol{\sigma}, \boldsymbol{\varepsilon}$, are used for second-order tensors. Fourth-order tensors are designated by upper case blackboard bold symbols, e.g., $\mathbb{A}, \mathbb{B}$. Higher-order tensors are denoted as $\mathbb{A}_{\langle r\rangle}$, where $\langle r\rangle$ represent the tensor order r. The dyadic / tensor product is denoted by $\otimes$. An orthonormal basis $\{\boldsymbol{b}_1, \boldsymbol{b}_2, \boldsymbol{b}_3\} = \{\boldsymbol{b}_i\}$ of the physical three dimensional Euclidean space is used throughout this work in order to represent all tensors, i.e., we use the representations $\boldsymbol{a} = \sum_{i=1}^3 a_i \boldsymbol{b}_i$, $\boldsymbol{A} = \sum_{i,j=1}^3 A_{ij} \boldsymbol{b}_{ij}$ or $\mathbb{A} = \sum_{i,j,k,l=1}^3 A_{ijkl} \boldsymbol{b}_{ijkl}$, where a_i, A_{ij}, A_{ijkl} denote the corresponding tensor components and $\boldsymbol{b}_i \otimes \boldsymbol{b}_j = \boldsymbol{b}_{ij}$ is abbreviated for better readability. Minor symmetric fourth-order tensors fulfill $A_{ijkl} = A_{jikl} = A_{ijlk}$, while major symmetric fourth-order tensors fulfill $A_{ijkl} = A_{klij}$. The linear map of a first-order tensor over a second-order tensor is denoted as $\boldsymbol{A}\boldsymbol{x} = \sum_{i,j=1}^3 A_{ij} x_j \boldsymbol{b}_i$, while higher-order linear maps are denoted by $\mathbb{A}[\boldsymbol{B}] = \sum_{i,j,k,l=1}^3 A_{ijkl} B_{kl} \boldsymbol{b}_{ij}$ and $\mathbb{A}_{\langle 6\rangle}[\mathbb{B}_{\langle 4\rangle}] = \sum_{i,j,k,l,m,n=1}^3 A_{ijklmn} B_{klmn} \boldsymbol{b}_{ij}$. The composition of two equal order tensors delivering a tensor of equal order is simply symbolized by $\mathbb{A}_{\langle r\rangle}\mathbb{B}_{\langle r\rangle}$, e.g., $\boldsymbol{A}\boldsymbol{B} = \sum_{i,j,k=1}^3 A_{ij} B_{jk} \boldsymbol{b}_{ik}$ maps as $(\boldsymbol{A}\boldsymbol{B})\boldsymbol{x} = \boldsymbol{A}(\boldsymbol{B}\boldsymbol{x})$ and $\mathbb{A}\mathbb{B} = \sum_{i,j,k,l,m,n=1}^3 A_{ijkl} B_{klmn} \boldsymbol{b}_{ijmn}$ maps as $(\mathbb{A}\mathbb{B})[\boldsymbol{C}] = \mathbb{A}[\mathbb{B}[\boldsymbol{C}]]$. Composition of tensors of different orders is not needed in this work. The scalar product is denoted as $\mathbb{A}_{\langle r\rangle} \cdot \mathbb{B}_{\langle r\rangle} = \sum_{i_1,\ldots,i_r=1}^3 A_{i_1\ldots i_r} B_{i_1\ldots i_r}$. The Frobenius norm is defined as $\|\mathbb{A}_{\langle r\rangle}\| = \sqrt{\mathbb{A}_{\langle r\rangle} \cdot \mathbb{A}_{\langle r\rangle}}$. The Rayleigh product is denoted by $\boldsymbol{A} \star \mathbb{B}_{\langle r\rangle} = \sum_{i_1,\ldots,i_r=1}^3 B_{i_1\ldots i_r} (\boldsymbol{A}\boldsymbol{b}_{i_1}) \otimes \ldots \otimes (\boldsymbol{A}\boldsymbol{b}_{i_r})$. Hereby, the Rayleigh product of a zeroth-order tensor (a scalar) equals the zeroth-order tensor itself, i.e., $\boldsymbol{A} \star W = W$. The identity on vectors and the permutation tensor are denoted by $\boldsymbol{I} = \sum_{i,j=1}^3 \delta_{ij} \boldsymbol{b}_{ij}$ and $\boldsymbol{\epsilon} = \sum_{i,j,k=1}^3 \varepsilon_{ijk} \boldsymbol{b}_{ijk}$, respectively, where δ_{ij} and ε_{ijk} denote the Kronecker delta and the permutation symbol, respectively. The basic fourth-order tensors $\mathbb{I} = \sum_{i,j,k,l=1}^3 \delta_{ik}\delta_{jl} \boldsymbol{b}_{ijkl}$ (identity on second-order tensors), $\mathbb{I}^S = \sum_{i,j,k,l=1}^3 \frac{1}{2}(\delta_{ik}\delta_{jl} + \delta_{il}\delta_{jk}) \boldsymbol{b}_{ijkl}$ (identity on symmetric second-order tensors), $\mathbb{P}_1 = \frac{1}{3}\boldsymbol{I} \otimes \boldsymbol{I}$ (identity on isotropic second-order tensors), $\mathbb{P}_2 = \mathbb{I}^S - \mathbb{P}_1$ (identity on symmetric and traceless second-order tensors) and $\mathbb{P}_3 = \mathbb{I} - \mathbb{I}^S$ (identity on skewed second-order tensors) are used throughout this work. In this work, at several points scalar quantities, say f, will depend on several tensor valued variables of different orders, e.g., $f = f(\boldsymbol{A}, x) = \frac{1}{2}\boldsymbol{A} \cdot \mathbb{B}[\boldsymbol{A}] - \boldsymbol{A} \cdot (x\boldsymbol{C}) - \frac{1}{2}dx^2$. For these kinds of scalar quantities, a supertensor notation following vector matrix notation is used. The supervector $\underline{A} = [\boldsymbol{A}, x]^\mathsf{T}$ with components $(\underline{A})_i$ with $i = 1, \ldots, n$ represents the n degrees of freedom of the considered variables and the supertensor $\underline{\underline{B}}$ with matrix components $(\underline{\underline{B}})_{ij}$ is constructed based on $\mathbb{B}, (-\boldsymbol{C})$ and $(-d)$ such that $f = \frac{1}{2}\underline{A}^\mathsf{T}\underline{\underline{B}}\,\underline{A} = \frac{1}{2}\sum_{i,j=1}^{n}(\underline{A})_i(\underline{\underline{B}})_{ij}(\underline{A})_j$ holds, where $\underline{A}^\mathsf{T}$ denotes the standard transposition.

A list of the frequently used symbols and acronyms is given at the end of this document.

Chapter 2

Calculus of orientations

2.1 Overview

In this chapter, the set of orientations $SO(3)$ and the CODF $f(\boldsymbol{Q})$ for $\boldsymbol{Q} \in SO(3)$ are discussed. Fourier expansions for central and general CODFs depending on texture eigenvalues and texture coefficients are discussed. Furthermore, the convex set of all possible texture eigenvalues and texture coefficients is sketched. An orientation dependent real tensor valued quantity $\mathbb{D}_{\langle r \rangle}(\boldsymbol{Q}) = \boldsymbol{Q} \star \tilde{\mathbb{D}}_{\langle r \rangle}$ of r-th-order, with constant reference single crystal material property $\tilde{\mathbb{D}}_{\langle r \rangle}$, and the orientation average of $\mathbb{D}_{\langle r \rangle}(\boldsymbol{Q})$ based on $f(\boldsymbol{Q})$ over $SO(3)$ are illustrated in this chapter. Quantities corresponding to the reference single crystal behavior will be denoted by a tilde, as $\tilde{\mathbb{D}}_{\langle r \rangle}$. The harmonic decomposition of the single crystal tensor valued quantity $\tilde{\mathbb{D}}_{\langle r \rangle}$ is illustrated in order to examine the orientation average of $\mathbb{D}_{\langle r \rangle}(\boldsymbol{Q})$ over $SO(3)$ with consideration of the tensorial Fourier expansion of $f(\boldsymbol{Q})$. Based on the harmonic decomposition, the orientation average is given in terms of the relevant influence of the CODF, i.e., in terms of texture eigenvalues and texture coefficients up to the respective tensor order. This representation gives access to a *finite and low dimensional* parametrization of the orientation averages depending on finite-dimensional variables belonging to convex sets. These parametrizations will be used in following chapters for the representation of bounds and approximations of linear elastic properties depending on the respective set of texture variables (texture eigenvalues and texture coefficients) for optimization of mechanical properties over the respective convex sets.

2.2 Orientations

2.2.1 The manifold $SO(3)$

Definition. The set $SO(3)$ is the manifold of special orthogonal second-order tensors (corresponding to proper rotations) in the 3-dimensional real-valued space of first-order tensors (vectors). Further, $SO(3)$ is a subset of the set of orthogonal (unitary) second-order tensors $Orth$, defined by

$$Orth = \{\boldsymbol{R} \mid (\boldsymbol{R}\boldsymbol{x}) \cdot (\boldsymbol{R}\boldsymbol{y}) = \boldsymbol{x} \cdot \boldsymbol{y} \, \forall \boldsymbol{x}, \boldsymbol{y} \in V\} \, . \tag{2.1}$$

The vector space V in this work is the 3-dimensional Euclidean space of first-order tensors, for which an orthonormal basis $\{\boldsymbol{b}_1, \boldsymbol{b}_2, \boldsymbol{b}_3\}$ will be used. For $\boldsymbol{R} \in Orth$, $\boldsymbol{R}^{-1} = \boldsymbol{R}^{\mathsf{T}}$ and $\det(\boldsymbol{R}) = \pm 1$ hold. The set of orientations $SO(3)$ is defined as

$$SO(3) = \{\boldsymbol{Q} \mid \boldsymbol{Q} \in Orth \wedge \det(\boldsymbol{Q}) = +1\} \, . \tag{2.2}$$

Invariant integration over $SO(3)$ with integration variable $\boldsymbol{Q}$ is denoted as

$$1 = \int_{SO(3)} 1 \mathrm{d}Q(\boldsymbol{Q}) = \int_{SO(3)} \mathrm{d}Q(\boldsymbol{R}_1 \boldsymbol{Q} \boldsymbol{R}_2) \, , \, \forall \boldsymbol{R}_1, \boldsymbol{R}_2 \in SO(3) \tag{2.3}$$

where $\mathrm{d}Q$ represents the volume element ensuring an invariant integration over $SO(3)$, see, e.g., Gel'fand et al. (1963) or Morawiec (2004). For the sake of brevity, the notation $\mathrm{d}Q = \mathrm{d}Q(\boldsymbol{Q})$ will be used throughout the manuscript.

The scalar product of scalar functions, say $f_1(\boldsymbol{Q})$ and $f_2(\boldsymbol{Q})$, over $SO(3)$ is defined as

$$\langle f_1, f_2 \rangle = \int\limits_{SO(3)} f_1(\boldsymbol{Q}) f_2^*(\boldsymbol{Q}) \mathrm{d}Q \,, \tag{2.4}$$

where $f_2^*(\boldsymbol{Q})$ denotes complex conjugation. A square integrable function $f(\boldsymbol{Q})$ over $SO(3)$ will be addressed as $f(\boldsymbol{Q}) \in L^2(SO(3))$.

In this work the symbol $\boldsymbol{Q}$ will be used in order to denote elements of $SO(3)$, in order to abbreviate $\boldsymbol{Q} \in SO(3)$ and keep the text readable. Further, it is shortly stressed out that orientations $\boldsymbol{Q}$ are defined symbolically, i.e., independently of any chosen representation basis of the underlying vector space V. An orientation $\boldsymbol{Q}$ is treated in this work as a second-order tensor and can be represented by $\boldsymbol{Q} = \sum_{i,j=1}^{3} Q_{ij} \boldsymbol{b}_{ij}$ with respect to a constant orthonormal basis $\{\boldsymbol{b}_i\}$ of the underlying vector space V. For example, a vector $\boldsymbol{x} = \sum_{i=1}^{3} x_i \boldsymbol{b}_i \in V$ is reoriented by an orientation $\boldsymbol{Q}$. This delivers a new vector $\boldsymbol{y} = \boldsymbol{Q}\boldsymbol{x}$, represented as $\boldsymbol{y} = \boldsymbol{Q}\boldsymbol{x} = \sum_{i,j=1}^{3} Q_{ij} x_j \boldsymbol{b}_i = \sum_{i=1}^{3} y_i \boldsymbol{b}_i$ with respect to the same basis $\{\boldsymbol{b}_i\}$. The tensor components Q_{ij} correspond, therefore, to the classical rotation matrix components. In order to be completely clear and avoid any misunderstanding, this will be explicitly addressed with the parametrization in terms of rotation axis and angle, and with the parametrization in terms of Euler angles (see following sections).

2.2.2 Symmetry transformations and isotropic tensors

Symmetry transformations and groups. A tensor $\tilde{\mathbb{D}}_{\langle r \rangle}$ of r-th-order (e.g., the single crystal stiffness of a material) is said to be invariant under the transformation $\boldsymbol{S} \in Orth$ if $\boldsymbol{S} \star \tilde{\mathbb{D}}_{\langle r \rangle} = \tilde{\mathbb{D}}_{\langle r \rangle}$ holds. The symmetry group $S \subset Orth$ of a tensorial quantity $\tilde{\mathbb{D}}_{\langle r \rangle}$ is the collection of all symmetry transformations, i.e.,

$$\boldsymbol{S} \star \tilde{\mathbb{D}}_{\langle r \rangle} = \tilde{\mathbb{D}}_{\langle r \rangle} \quad \forall \boldsymbol{S} \in S \subset Orth \,. \tag{2.5}$$

The group symmetrization of a tensor $\tilde{\mathbb{D}}_{\langle r \rangle}$ with respect to a symmetry group S^{sym} with finite number n^{sym} of symmetry transformations is defined in this work as

$$\text{gs}^{\text{sym}}(\tilde{\mathbb{D}}_{\langle r \rangle}) = \frac{1}{n^{\text{sym}}} \sum_{\alpha=1}^{n^{\text{sym}}} \boldsymbol{S}_\alpha \star \tilde{\mathbb{D}}_{\langle r \rangle} \,, \quad \boldsymbol{S}_\alpha \in S^{\text{sym}} \, \forall \alpha \,. \tag{2.6}$$

Isotropic tensors. Isotropic tensors $\mathbb{B}^I_{\langle r \rangle}$ of order r are defined as

$$\boldsymbol{S} \star \mathbb{B}^I_{\langle r \rangle} = \mathbb{B}^I_{\langle r \rangle} \quad \forall \boldsymbol{S} \in Orth \,. \tag{2.7}$$

The components of isotropic tensors are identical in all orthonormal coordinate systems.

2.2.3 Parametrization of orientations

Rotation axis and angle parametrization. An orientation $\boldsymbol{Q}$ can be parametrized with a normalized rotation axis $\boldsymbol{n}$ and a rotation angle $\omega \in [0, \pi]$ as follows

$$\begin{gathered} \boldsymbol{Q} = \boldsymbol{N}_0 + \boldsymbol{N}_1 + \boldsymbol{N}_2 \,, \\ \boldsymbol{N}_0 = \cos(\omega)\boldsymbol{I} \,, \ \boldsymbol{N}_1 = -\sin(\omega)\boldsymbol{\epsilon}[\boldsymbol{n}] \,, \ \boldsymbol{N}_2 = (1 - \cos(\omega))\boldsymbol{n}^{\otimes 2} \,. \end{gathered} \tag{2.8}$$

For example, the orientation $\boldsymbol{Q} = \boldsymbol{Q}(\boldsymbol{n}, \omega)$, which rotates vectors around the axis $\boldsymbol{n} = \boldsymbol{b}_3$, is represented with respect to the basis $\{\boldsymbol{b}_i\}$ as

$$\boldsymbol{Q}(\boldsymbol{b}_3, \omega) = \sum_{i,j=1}^{3} Q_{ij}^{3,\omega} \boldsymbol{b}_{ij} \,, \quad Q_{ij}^{3,\omega} = \begin{pmatrix} c_\omega & -s_\omega & 0 \\ s_\omega & c_\omega & 0 \\ 0 & 0 & 1 \end{pmatrix} , \tag{2.9}$$

where $\cos(\omega) = c_\omega$ and $\sin(\omega) = s_\omega$ have been abbreviated for future expressions. This delivers for $\boldsymbol{y} = \boldsymbol{Q}\boldsymbol{x}$ with the linear combinations $\boldsymbol{x} = \sum_{i=1}^{3} x_i \boldsymbol{b}_i$ and $\boldsymbol{y} = \sum_{i=1}^{3} y_i \boldsymbol{b}_i$ the component relations

$$y_i = \sum_{j=1}^{3} Q_{ij}^{3,\omega} x_j = \begin{pmatrix} x_1 c_\omega - x_2 s_\omega \\ x_1 s_\omega + x_2 c_\omega \\ x_3 \end{pmatrix} . \tag{2.10}$$

The reader should consider this example to see, that the tensor representations of orientations used in this work might be different of the ones used in literature, see, e.g., Bunge (1982), where the inverse/transposed relations are used for the matrix components.

For a given orientation $\boldsymbol{Q}$, the rotation angle $\omega = \omega(\boldsymbol{Q})$ is obtained by

$$\omega = \arccos\left(\frac{1}{2}\left(\mathrm{tr}\left(\boldsymbol{Q}\right) - 1\right)\right) . \tag{2.11}$$

If this parametrization is used, then $\mathrm{d}Q$ is given as, see, for example, Morawiec (2004),

$$\mathrm{d}Q = \mathrm{d}n \; s(\omega)\mathrm{d}\omega \; , \quad s(\omega) = \frac{\sin^2(\omega/2)}{2\pi^2} \; , \tag{2.12}$$

where $\mathrm{d}n$ denotes the surface element of the unit sphere, denoted in this work as S_2. If the rotation axis is parametrized with spherical coordinates

$$\boldsymbol{n} = \sin(\theta)\cos(\phi)\boldsymbol{b}_1 + \sin(\theta)\sin(\phi)\boldsymbol{b}_2 + \cos(\theta)\boldsymbol{b}_3 \; , \tag{2.13}$$

with $\theta \in [0, \pi]$ and $\phi \in [0, 2\pi)$, then $\mathrm{d}n$ is given as

$$\mathrm{d}n = \sin(\theta)\mathrm{d}\phi\mathrm{d}\theta \; . \tag{2.14}$$

The integration over the unit sphere S_2 is performed as

$$\int_{S_2} 1 \mathrm{d}n = \int_{\theta=0}^{\pi} \int_{\phi=0}^{2\pi} 1 \sin(\theta) \mathrm{d}\phi \mathrm{d}\theta = 4\pi \; . \tag{2.15}$$

The fulfillment of the normalization condition can be verified as

$$\int_{SO(3)} 1 \mathrm{d}Q = \int_{\omega=0}^{\pi} \int_{S_2} 1 \mathrm{d}n \; s(\omega) \mathrm{d}\omega = 1 \; . \tag{2.16}$$

Example orientation symmetry groups. Based on the parametrization Eq. (2.8) for an orientation $\boldsymbol{Q} = \boldsymbol{Q}(\boldsymbol{n}, \omega)$, several symmetry groups $S^i \subset SO(3) \subset Orth$ of solid physical quantities with respect to an orthonormal vector basis $\{\boldsymbol{b}_1, \boldsymbol{b}_2, \boldsymbol{b}_3\}$ can be described. For second- and fourth-order physical tensorial quantities, the groups

$$\begin{aligned} S^{\mathrm{ort}} &= \{\boldsymbol{S} \mid \boldsymbol{S} = \boldsymbol{Q}(\boldsymbol{b}_i, j\,\pi), i \in \{1,2,3\}, j \in \{0,1\}\} \; , \\ S^{\mathrm{hex}} &= \{\boldsymbol{S} \mid \boldsymbol{S} = \boldsymbol{Q}(\boldsymbol{b}_3, i\,2\pi/6) \; , \; i \in \{0,1,\ldots,5\}\} \; , \\ S^{\mathrm{cub}} &= \{\boldsymbol{S} \mid \boldsymbol{S} = \boldsymbol{Q}(\boldsymbol{n}_i, \omega_i), (\boldsymbol{n}_i, \omega_i) \in S^{\mathrm{cub}}_{n\omega}\} \; , \end{aligned} \tag{2.17}$$

with

$$\begin{aligned} S^{\mathrm{cub}}_{n\omega} = \; & \{(\boldsymbol{b}_i, j2\pi/4), i \in \{1,2,3\}, j \in \{0,1,2,3\}\} \\ & \cup\{((\boldsymbol{b}_i + (-1)^k \boldsymbol{b}_j)/\sqrt{2}, \pi), i \neq j, i,j \in \{1,2,3\} \\ & \quad , k \in \{1,2\}\} \\ & \cup\{((\boldsymbol{b}_i + (-1)^l \boldsymbol{b}_j + (-1)^m \boldsymbol{b}_k)/\sqrt{3}, n\,2\pi/3)\} \\ & \quad , i \neq j \neq k, i,j,k \in \{1,2,3\}, l,m,n \in \{1,2\}\} \end{aligned} \tag{2.18}$$

represent the classical rotational symmetry groups for orthotropic (with respect to $\{\boldsymbol{b}_i\}$), hexagonal (with respect to $\boldsymbol{b}_3$) and cubic (with respect to $\{\boldsymbol{b}_i\}$) solid materials.

Example discretization of $SO(3)$. In this work, the parametrizations Eq. (2.8) of $\boldsymbol{Q} = \boldsymbol{Q}(\boldsymbol{n}, \omega)$ and Eq. (2.13) of the rotation axis $\boldsymbol{n} = \boldsymbol{n}(\theta, \phi)$ are used to demonstrate several concepts over $SO(3)$. Based on these representations, the simplest discretization of $SO(3)$ is the following

$$\begin{aligned} Q_n &= \{\boldsymbol{R} \mid \boldsymbol{R} = \boldsymbol{Q}(\boldsymbol{n}(\theta, \phi), \omega), \omega \in [\pi]_n, \theta \in [\pi]_n, \phi \in [2\pi)_n\} \, , \\ [\pi]_n &= \{0, \pi/n, \ldots, \pi\} \, , \; [2\pi)_n = \{0, 2\pi/n, \ldots, 2\pi(1 - 1/n)\} \, . \end{aligned} \tag{2.19}$$

Naturally, this discretization is not equidistant in $SO(3)$ and other much better discretizations of $SO(3)$ can be used, e.g., the approximately equidistant distribution of orientations, see, e.g., Helming (1997) or Rosca et al. (2014). However, in this chapter, it is not essential to investigate different discretizations of $SO(3)$, but to illustrate different concepts using a discretization. The discretization Q_n of Eq. (2.19) has been chosen in the present work, such that the reader can reproduce the illustrated concepts of this chapter without unnecessary complications.

Alternatives. Alternative parametrizations can also be considered, e.g., different conventions for the Euler angles or the Rodrigues parameters of a rotation connected to a corresponding quaternion, see, e.g., Bunge (1982), Schaeben (1990), Morawiec (2004), Mardia and Jupp (2008). In order to clearly address differences with parametrizations used in literature, the Euler angle parametrization of an orientation $\boldsymbol{Q}$ is shortly sketched. First, we consider the orientation $\boldsymbol{Q}^{3,\varphi_1}$ which rotates vectors around the axis $\boldsymbol{n} = \boldsymbol{b}_3$ with rotation angle $\omega = \varphi_1$. This orientation can be represented with respect to $\{\boldsymbol{b}_i\}$ as $\boldsymbol{Q}^{3,\varphi_1} = \sum_{i,j=1}^{3} Q_{ij}^{3,\varphi_1} \boldsymbol{b}_{ij}$ with Q_{ij}^{3,φ_1} taken from Eq. (2.9). The basis vector $\boldsymbol{b}_i$ are mapped to the vectors $\boldsymbol{b}'_i = \boldsymbol{Q}^{3,\varphi_1} \boldsymbol{b}_i = \sum_{j=1}^{3} Q_{ji}^{3,\varphi_1} \boldsymbol{b}_j$, where $\{\boldsymbol{b}'_i\}$ forms an alternative orthonormal basis and $\boldsymbol{b}'_i \cdot \boldsymbol{b}_j = Q_{ji}^{3,\varphi_1}$ holds. For the following reasoning, the alternative representation $\boldsymbol{Q}^{3,\varphi_1} = \sum_{i=1}^{3} \boldsymbol{b}'_i \otimes \boldsymbol{b}_i$ is useful. Now, we consider a second orientation $\boldsymbol{Q}^{1,\Phi}$ which rotates vectors around the axis

$\boldsymbol{n} = \boldsymbol{b}_1'$ with the angle $\omega = \Phi$. This second orientation can be represented with respect to $\{\boldsymbol{b}_i'\}$ as

$$\boldsymbol{Q}^{1,\Phi} = \boldsymbol{Q}(\boldsymbol{b}_1', \Phi) = \sum_{i,j=1}^{3} Q_{ij}^{1,\Phi} \boldsymbol{b}_{ij}' \, , \; Q_{ij}^{1,\Phi} = \begin{pmatrix} 1 & 0 & 0 \\ 0 & c_\Phi & -s_\Phi \\ 0 & s_\Phi & c_\Phi \end{pmatrix} . \quad (2.20)$$

The vectors $\boldsymbol{b}_i'$ are mapped through $\boldsymbol{Q}^{1,\Phi}$ to the orthonormal vectors $\boldsymbol{b}_i'' = \boldsymbol{Q}^{1,\Phi}\boldsymbol{b}_i' = \sum_{j=1}^{3} Q_{ji}^{1,\Phi}\boldsymbol{b}_j'$, and the orientation may be expressed as $\boldsymbol{Q}^{1,\Phi} = \sum_{i=1}^{3} \boldsymbol{b}_i'' \otimes \boldsymbol{b}_i'$. Finally, we consider a third orientation $\boldsymbol{Q}^{3,\varphi_2}$ rotating vectors around the axis $\boldsymbol{n} = \boldsymbol{b}_3''$ by the angle $\omega = \varphi_2$. The orientation is represented with respect to $\{\boldsymbol{b}_i''\}$ as $\boldsymbol{Q}^{3,\varphi_2} = \sum_{i,j=1}^{3} Q_{ij}^{3,\varphi_2}\boldsymbol{b}_{ij}''$, where Q_{ij}^{3,φ_2} is taken from Eq. (2.9). The vectors $\boldsymbol{b}_i''$ are mapped through $\boldsymbol{Q}^{3,\varphi_2}$ to the vectors $\boldsymbol{b}_i''' = \boldsymbol{Q}^{3,\varphi_2}\boldsymbol{b}_i'' = \sum_{j=1}^{3} Q_{ji}^{3,\varphi_2}\boldsymbol{b}_j''$, and the orientation may be expressed as $\boldsymbol{Q}^{3,\varphi_2} = \sum_{i=1}^{3} \boldsymbol{b}_i''' \otimes \boldsymbol{b}_i''$. Now, the sequential application of these three orientations can be computed and represented with respect to $\{\boldsymbol{b}_i\}$ as

$$\begin{aligned} \boldsymbol{Q}^{\varphi_1\Phi\varphi_2} &= \boldsymbol{Q}^{3,\varphi_2}\boldsymbol{Q}^{1,\Phi}\boldsymbol{Q}^{3,\varphi_1} = \sum_{j=1}^{3} \boldsymbol{b}_j''' \otimes \boldsymbol{b}_j \\ &= \sum_{i,j,k,l=1}^{3} Q_{kj}^{3,\varphi_2} Q_{lk}^{1,\Phi} Q_{il}^{3,\varphi_1} \boldsymbol{b}_{ij} = \sum_{i,j=1}^{3} Q_{ij}^{\varphi_1\Phi\varphi_2} \boldsymbol{b}_{ij} \end{aligned} \quad (2.21)$$

with the components

$$\begin{aligned} Q_{ij}^{\varphi_1\Phi\varphi_2} &= \sum_{k,l=1}^{3} Q_{il}^{3,\varphi_1} Q_{lk}^{1,\Phi} Q_{kj}^{3,\varphi_2} \\ &= \begin{pmatrix} c_{\varphi_1}c_{\varphi_2} - c_\Phi s_{\varphi_1}s_{\varphi_2} & -c_\Phi c_{\varphi_2}s_{\varphi_1} - c_{\varphi_1}s_{\varphi_2} & s_\Phi s_{\varphi_1} \\ c_{\varphi_2}s_{\varphi_1} + c_\Phi c_{\varphi_1}s_{\varphi_2} & c_\Phi c_{\varphi_1}c_{\varphi_2} - s_{\varphi_1}s_{\varphi_2} & -c_{\varphi_1}s_\Phi \\ s_\Phi s_{\varphi_2} & c_{\varphi_2}s_\Phi & c_\Phi \end{pmatrix} . \end{aligned} \quad (2.22)$$

As stated before, the reader should keep in mind, that in some literature inverse/transposed relations are used (cf. Bunge (1982), p.21, Eq. (2.50)).

2.3 The crystallite orientation distribution function (CODF)

2.3.1 General properties

Definition. Polycrystals may show complex distributions of the orientation of the crystallographic axes of a considered material. The simplest description of a polycrystal is the consideration of the total volume of the material having an orientation $\boldsymbol{Q}$ referred to the total volume of the polycrystal, denoted as $\mathrm{d}v(\boldsymbol{Q})/v$. This allows to define an orientation density function $f(\boldsymbol{Q})$ as $\mathrm{d}v(\boldsymbol{Q})/v = f(\boldsymbol{Q})\mathrm{d}Q$, see Bunge (1982). The density function $f(\boldsymbol{Q})$ is sometimes referred to in the field of quantitative texture analysis of polycrystalline materials as the crystallite orientation distribution function (CODF), see Bunge (1982). Strictly speaking, $f(\boldsymbol{Q})$ is a density function, see, e.g., Morawiec and Pospiech (1992) or Schaeben and van den Boogaart (2003), not a distribution function, although this term is sometimes used in literature, see, e.g., Bunge (1982), Matthies et al. (1988), Morawiec (1989), Böhlke (2006), Man and Huang (2012) and Adams et al. (2013). The CODF reflects one-point statistical information of the polycrystal. Two-point statistical information of the polycrystal can be represented by an orientation correlation function, see, e.g., Morawiec and Pospiech (1992).
The CODF is non-negative and normalized, i.e.,

$$f(\boldsymbol{Q}) \geq 0 \quad \forall \boldsymbol{Q} \in SO(3) \,, \qquad \int\limits_{SO(3)} f(\boldsymbol{Q})\mathrm{d}Q = 1 \,. \tag{2.23}$$

If a polycrystalline region is reoriented homogeneously in space with an orientation $\boldsymbol{R}$, then the orientation of each single crystal region in the polycrystal is shifted from $\boldsymbol{Q}$ to $\boldsymbol{RQ}$. It is self-evident that the relative amounts of the orientations are kept the same, i.e.,

$$\boldsymbol{R} \star \int\limits_{SO(3)} f(\boldsymbol{Q})\boldsymbol{Q} \star \tilde{\mathbb{D}}_{\langle r \rangle} \mathrm{d}Q = \int\limits_{SO(3)} f(\boldsymbol{R}^{-1}\boldsymbol{Q})\boldsymbol{Q} \star \tilde{\mathbb{D}}_{\langle r \rangle} \mathrm{d}Q \quad \forall \boldsymbol{R} \in SO(3) \tag{2.24}$$

holds due to (2.3) for the general CODF.

The orientation average of the real tensor valued $\mathbb{D}_{\langle r \rangle}(\boldsymbol{Q}) = \boldsymbol{Q} \star \tilde{\mathbb{D}}_{\langle r \rangle}$ with the CODF $f(\boldsymbol{Q})$ may also be expressed based on Eq. (2.4) as

$$\langle f, \mathbb{D}_{\langle r \rangle} \rangle = \int\limits_{SO(3)} f(\boldsymbol{Q})\mathbb{D}_{\langle r \rangle}(\boldsymbol{Q}) \mathrm{d}Q \, . \tag{2.25}$$

This notation will be used throughout this chapter.

Special cases. A single crystal with orientation $\hat{\boldsymbol{Q}}$ is described by the Dirac-distribution $f(\boldsymbol{Q}) = \delta(\hat{\boldsymbol{Q}}, \boldsymbol{Q})$ which acts on a function $h(\boldsymbol{Q})$ as

$$\int\limits_{SO(3)} \delta(\hat{\boldsymbol{Q}}, \boldsymbol{Q}) h(\boldsymbol{Q}) \mathrm{d}Q = h(\hat{\boldsymbol{Q}}) \, . \tag{2.26}$$

A uniform distribution of orientations is described by $f(\boldsymbol{Q}) = 1$, which is referred to as the crystallographically isotropic state.

Symmetry groups. The CODF reflects the microscopic symmetry group S_{micro} of the considered material quantity, i.e.,

$$\boldsymbol{S} \star \tilde{\mathbb{D}}_{\langle r \rangle} = \tilde{\mathbb{D}}_{\langle r \rangle} \; \forall \boldsymbol{S} \in S_{\text{micro}} \; \Rightarrow \; f(\boldsymbol{QS}) = f(\boldsymbol{Q}) \; \forall \boldsymbol{S} \in S_{\text{micro}} \, . \tag{2.27}$$

Further, the macroscopic symmetry group S_{macro} of the material region is also reflected by the CODF,

$$\boldsymbol{S} \star \langle f, \mathbb{D}_{\langle r \rangle} \rangle = \langle f, \mathbb{D}_{\langle r \rangle} \rangle \; \forall \boldsymbol{S} \in S_{\text{macro}} \;\Rightarrow\; f(\boldsymbol{SQ}) = f(\boldsymbol{Q}) \; \forall \boldsymbol{S} \in S_{\text{macro}} \,. \tag{2.28}$$

2.3.2 Central CODF (CCODF)

Orientation distance. A CODF is referred to as central in this work, if it depends only on the distance $\hat{\omega}$ between a central orientation $\hat{\boldsymbol{Q}}$ and the considered orientation $\boldsymbol{Q}$, where $\hat{\omega}$ is defined as

$$\hat{\omega} = \arccos\left(\frac{1}{2}\left(\operatorname{tr}\left(\boldsymbol{Q}\hat{\boldsymbol{Q}}^{-1}\right) - 1\right)\right) \in [0, \pi] \,. \tag{2.29}$$

In a first attempt, a central CODF (CCODF) $g(\hat{\omega})$ might be interpreted as a normal distribution in $SO(3)$, see, e.g., Matthies et al. (1988); however, it should be noted that this interpretation is to be considered carefully, as discussed in Schaeben (1992).

If the CCODF is considered centered around the identity, i.e., $\hat{\boldsymbol{Q}} = \boldsymbol{I}$, then the orientation distance $\hat{\omega}$ (Eq. (2.29)) is equal to the rotation angle ω (Eq. (2.11)), and the CCODF $g(\omega)$ describes the distribution as a function of the rotation angle independently of the rotation axis.

Example CCODFs. CCODFs have been used in the field of quantitative texture analysis to approximate complex CODFs. Some model CCODFs are the von Mises-Fisher CCODF

$$\begin{aligned} g^{MF}(\hat{\omega}) &= \frac{\exp(S\cos(\hat{\omega}))}{I_0(S) - I_1(S)} \,, \quad S \geq 0 \,, \\ I_n(S) &= \frac{1}{\pi}\int_0^{\pi} \exp(S\cos(n\omega))\cos(n\omega)\mathrm{d}\omega \,, \end{aligned} \tag{2.30}$$

and the Lorentzian CCODF (also referred to as Cauchy distribution in probability)

$$g^L(\hat{\omega}) = (1 - S^2)\frac{(1+S^2)^2 + 4S^2\cos^2(\hat{\omega}/2)}{((1+S^2)^2 - 4S^2\cos^2(\hat{\omega}/2))^2}\,, \quad 0 \leq S < 1\,, \quad (2.31)$$

see, e.g., Matthies et al. (1988), Schaeben (1990), Morawiec (2004). The parameter S represents a concentration parameter. For a vanishing parameter, i.e., $S = 0$, the model CCODF $g^X(\hat{\omega})$ transitions into the uniform CODF. An increasing parameter sharpens the model CCODF around the central orientation $\hat{\boldsymbol{Q}}$ indicating the tendency towards a single crystal.

Approximation of CODFs. Depending on the observed material deformation during processing of polycrystalline materials, the CODF of the investigated material may or may not be well representable by a convex combination of n_c model CCODFs $g_\alpha(\hat{\omega}_\alpha)$ with respective central orientations $\hat{\boldsymbol{Q}}_\alpha$

$$\begin{aligned} f(\boldsymbol{Q}) &= \sum_{\alpha=1}^{n_c} f_\alpha g_\alpha(\hat{\omega}_\alpha)\,, \quad f_\alpha \geq 0 \forall \alpha\,, \quad \sum_{\alpha=1}^{n_c} f_\alpha = 1\,, \\ \hat{\omega}_\alpha &= \hat{\omega}\Big|_{\hat{\boldsymbol{Q}}=\hat{\boldsymbol{Q}}_\alpha} = \arccos\left(\frac{1}{2}\left(\mathrm{tr}\left(\boldsymbol{Q}\hat{\boldsymbol{Q}}_\alpha^{-1}\right) - 1\right)\right) \end{aligned} \quad (2.32)$$

The ansatz of Eq. (2.32) is known in the field of quantitative texture analysis as texture component method, for small n_c. For large n_c, the ansatz Eq. (2.32) may be considered as an approximation by convex combinations of central / radially symmetric basis functions. The different model functions $g(\hat{\omega}_\alpha)$ are referred to as components and algorithms to determine the weight f_α, the central orientation and possible concentration parameter of each component are presented in literature, see, e.g., Wassermann and Grewen (1962), Lücke et al. (1981), Schaeben (1996), Helming (1998) and Böhlke et al. (2006). However, the question if a complex CODF is representable by a convex combination of CCODFs is not at the focus of this section. The question approached in this section is: What is the *relevant* influence of texture on the orientation average of a tensorial quantity *if* the general CODF is approximated as a convex combination of CCODFs? Does the complete CCODF influence the orientation average of the tensorial quantity *or only a spectral part of it*? This will be analyzed in the following sections.

Fourier expansion of a CCODF in terms of texture eigenvalues. If a CCODF centered around the identity, $g(\omega)$, is assumed square integrable, $g(\omega) \in L^2(SO(3))$, then its Fourier expansion exists, which is denoted in this work as

$$g(\omega) = \sum_{\alpha=0}^{\infty} (1+2\alpha)^2 \lambda_\alpha \mu_\alpha(\omega) \,. \tag{2.33}$$

The ansatz functions $\mu_\alpha(\omega)$ are constructed based on the Dirichlet kernels $D_\alpha(\omega)$, see Savyolova (1984) or Schaeben (1992),

$$\mu_\alpha(\omega) = \frac{D_\alpha(\omega)}{(1+2\alpha)} \,, \quad D_0(\omega) = 1 \,, \quad D_\alpha(\omega) = 1 + 2\sum_{k=1}^{\alpha} \cos(k\omega) \,. \tag{2.34}$$

The Dirichlet kernels build an orthonormal system with respect to $L^2(SO(3))$, while the ansatz functions build an orthogonal system, i.e.,

$$\int_{SO(3)} \mu_\alpha(\omega)\mu_\beta(\omega) \mathrm{d}Q = \begin{cases} \frac{1}{(1+2\alpha)^2} & \alpha = \beta \\ 0 & \text{else} \end{cases} \,. \tag{2.35}$$

The scalar Fourier coefficients λ_α are determined as

$$\lambda_0 = 1 \,, \quad \lambda_\alpha = \int_{SO(3)} g(\omega)\mu_\alpha(\omega) \mathrm{d}Q \,. \tag{2.36}$$

The Fourier coefficients λ_α will be referred to as texture eigenvalues in the present work, as in Lobos et al. (2017), for reasons to be discussed later in Section 2.5.2. Each texture eigenvalue is bounded from below and above as follows

$$\min_{\boldsymbol{Q}\in SO(3)} \mu_\alpha(\omega) \leq \lambda_\alpha \leq \max_{\boldsymbol{Q}\in SO(3)} \mu_\alpha(\omega) = 1\,, \tag{2.37}$$

where the minimum and maximum values are computed in the end only with respect to the rotation angle ω since the functions $\mu_\alpha(\omega)$ are independent of the rotation axis. The bounds of the texture eigenvalues are immediately visible by considering their definition as a simple convex combination of the values of the corresponding $\mu_\alpha(\omega)$ over $SO(3)$. The maximal value for all functions is found at the origin, meaning that the texture eigenvalues equal unity for the limiting case of the single crystal centered at $\hat{\boldsymbol{Q}} = \boldsymbol{I}$. For the isotropic case, i.e., $g(\omega) = 1$, all texture eigenvalues λ_α for $\alpha \geq 1$ vanish.

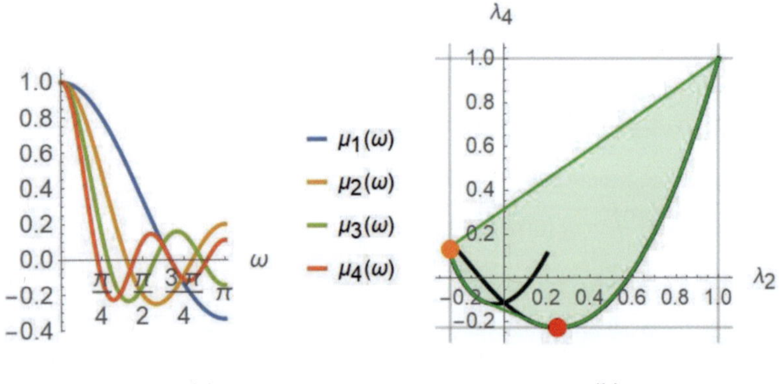

Figure 2.1: (a) Functions $\mu_\alpha(\omega)$ for $\alpha \in \{1, 2, 3, 4\}$; (b) Set of all possible texture eigenvalues (λ_2, λ_4)

For the present work, the texture eigenvalues λ_2 and λ_4 are important in the upcoming orientation averages in Chapter 3 and are discussed briefly. First, consider the functions $\mu_\alpha(\omega), \alpha \in \{1, 2, 3, 4\}$, depicted in Fig. 2.1a. The maximum of all functions is found at the origin. It is also visible that all $\mu_\alpha(\omega)$ can change their sign and that $|\mu_\alpha(\omega)| \leq 1$ holds. All these observations motivate the properties

1. that texture eigenvalues with negative sign exist,
2. that, due to the convex combination of the values of the functions $\mu_\alpha(\omega)$, $|\lambda_\alpha| \leq 1$ holds, and
3. that the set of all possible texture eigenvalues of order 2 and 4 is the convex hull of the curve described by the functions $\mu_\alpha(\omega)$.

These properties are easily seen, e.g., by examination of the vector

$$\begin{pmatrix} \lambda_2 \\ \lambda_4 \end{pmatrix} = \int_{SO(3)} g(\omega) \begin{pmatrix} \mu_2(\omega) \\ \mu_4(\omega) \end{pmatrix} \mathrm{d}Q \,. \tag{2.38}$$

Now consider the CCODF $g^\delta(\omega) = \delta(\omega - \omega_0)/(4\pi s(\omega_0))$ for $\omega_0 \in (0, \pi]$. This CCODF represents a CODF accepting only orientations $\boldsymbol{Q}$ with a distance equal to ω_0 from $\hat{\boldsymbol{Q}} = \boldsymbol{I}$. The artificial CCODF $g^\delta(\omega)$ might not be easily found in nature, but it fulfills the normalization condition and is non-negative. Therefore, $g^\delta(\omega)$ is admissible as a CCODF. Consideration of $g^\delta(\omega)$ in Eq. (2.38) delivers the value of the functions $\mu_\alpha(\omega)$ at $\omega = \omega_0$ and the corresponding point in the 2-dimensional real domain. For ω_0 minimizing $\mu_4(\omega)$, see Fig. 2.1a, $\mu_2(\omega)$ is positive and the red point depicted in Fig. 2.1b is obtained. Analogously, the orange point is obtained for ω_0 minimizing $\mu_2(\omega)$. The black line is obtained by considering all $\omega_0 \in (0, \pi]$, where the limit for $\omega_0 \to 0$ corresponds to the single crystal with $\lambda_\alpha = 1$. Now, any other CCODF $g(\omega)$ will make a convex combination of the points on the black line, meaning that, without further specification of the actual CCODF $g(\omega)$, all possible texture eigenvalues (λ_2, λ_4) lie in the convex hull of the curve. This set is depicted in green in Fig. 2.1b and will be addressed for the texture eigenvalues (λ_2, λ_4) as

$$\Lambda_{2,4} = \left\{ (\lambda_2, \lambda_4) \mid \begin{pmatrix} \lambda_2 \\ \lambda_4 \end{pmatrix} = \int_{SO(3)} g(\omega) \begin{pmatrix} \mu_2(\omega) \\ \mu_4(\omega) \end{pmatrix} \mathrm{d}Q \right\}, \tag{2.39}$$

where $g(\omega)$ denotes a general CCODF.

Naturally, this reasoning can be extended to all texture eigenvalues which belong to the infinite-dimensional convex set

$$\Lambda = \left\{ (\lambda_1, \lambda_2, ...) \mid \begin{pmatrix} \lambda_1 \\ \lambda_2 \\ \vdots \end{pmatrix} = \int_{SO(3)} g(\omega) \begin{pmatrix} \mu_1(\omega) \\ \mu_2(\omega) \\ \vdots \end{pmatrix} \mathrm{d}Q \right\}, \tag{2.40}$$

which is embedded in the rectangle described by Eq. (2.37). The set $\Lambda_{2,4}$ represents a projection of Λ and the gray lines depicted in Fig. 2.1b represent the corresponding bounds defined in Eq. (2.37).

2.3.3 Tensorial Fourier expansion (TFE)

The tensorial Fourier expansion. If the general CODF $f(\boldsymbol{Q})$ is assumed square integrable, then it may be represented by a Fourier expansion. In this work, following Guidi et al. (1992) and Lobos Fernández and Böhlke (2018), the space of harmonic tensors is used to generate real tensor valued functions for the representation of the CODF. A harmonic tensor $\mathbb{H}'_{\langle\alpha\rangle}$ of α-th tensor-order is a fully symmetric tensor fulfilling $\mathbb{H}'_{\langle\alpha\rangle}[\boldsymbol{I}] = \mathbb{O}_{\langle\alpha-2\rangle}$ (referred to as traceless). The term harmonic is referred to these tensors due to their connection to homogeneous polynomials with vanishing Laplacian. Harmonic tensors will be denoted by a prime, as, e.g., $\mathbb{H}'_{\langle\alpha\rangle}$. A α-th-order harmonic tensor has, in general, $1 + 2\alpha$ degrees of freedom. We consider for each α-th-order an orthonormal basis of harmonic tensors $\{\mathbb{H}'_{\langle\alpha\rangle\beta}\}$ (i.e., $\mathbb{H}'_{\langle\alpha\rangle\beta} \cdot \mathbb{H}'_{\langle\alpha\rangle\tilde{\beta}} = \delta_{\beta\tilde{\beta}}$) with $\beta = 1, \ldots, n'_\alpha \leq 1 + 2\alpha$. These basis harmonic tensors $\mathbb{H}'_{\langle\alpha\rangle\beta}$ will be denoted as reference tensors. The number n'_α of reference tensors of α-th-order will be discussed later based on the conditions implied by the microscopic properties of the CODF. The reference tensors are now used in order to generate the tensor functions

$$\mathbb{F}'_{\langle\alpha\rangle\beta}(\boldsymbol{Q}) = \boldsymbol{Q} \star \mathbb{H}'_{\langle\alpha\rangle\beta} \,, \quad \|\mathbb{F}'_{\langle\alpha\rangle\beta}(\boldsymbol{Q})\| = 1 \,, \tag{2.41}$$

which form an orthogonal system with respect to $SO(3)$

$$\int\limits_{SO(3)} \mathbb{F}'_{\langle\alpha\rangle\beta}(\boldsymbol{Q}) \otimes \mathbb{F}'_{\langle\tilde{\alpha}\rangle\tilde{\beta}}(\boldsymbol{Q}) \mathrm{d}Q = \begin{cases} \dfrac{\mathbb{I}'_{\langle 2\alpha\rangle}}{1 + 2\alpha} & \alpha = \tilde{\alpha} \wedge \beta = \tilde{\beta} \\ \mathbb{O}_{\langle\alpha+\tilde{\alpha}\rangle} & \text{else} \end{cases} \,, \tag{2.42}$$

where $\mathbb{I}'_{\langle 2\alpha\rangle}$ denotes the identity on harmonic α-th-order tensors. Based on the tensor functions given in Eq. (2.41), the tensorial Fourier expansion (TFE) of the CODF may be expressed as

$$f(\boldsymbol{Q}) = \sum_{\alpha=0}^{\infty} \sum_{\beta=1}^{n'_\alpha} (1 + 2\alpha) \mathbb{V}'_{\langle\alpha\rangle\beta} \cdot \mathbb{F}'_{\langle\alpha\rangle\beta}(\boldsymbol{Q}) \,, \tag{2.43}$$

see also, e.g., Guidi et al. (1992), Zheng and Fu (2001), Böhlke (2005), Kalisch and Bertram (2013) and Lobos Fernández and Böhlke (2018). For the TFE given in Eq. (2.43), the harmonic α-th-order tensors $\mathbb{V}'_{\langle\alpha\rangle\beta}$, referred to in the field of texture analysis as (tensorial) texture coefficients, are determined by

$$\mathbb{V}'_{\langle\alpha\rangle\beta} = \langle f, \mathbb{F}'_{\langle\alpha\rangle\beta} \rangle \,, \tag{2.44}$$

see Böhlke (2005; 2006). The zeroth-order reference tensor is defined as $\mathbb{H}'_{\langle 0\rangle} = 1$, implying $\mathbb{F}'_{\langle 0\rangle}(\boldsymbol{Q}) = 1$ and $\mathbb{V}'_{\langle 0\rangle} = 1$. Due to the properties of the CODF, the texture coefficients $\mathbb{V}'_{\langle\alpha\rangle\beta}$ might be immediately interpreted as the convex combination of single crystal states of normalized reference tensors. It follows that the norm of the texture coefficients is bounded by $\|\mathbb{V}'_{\langle\alpha\rangle\beta}\| \in [0, 1]$, where the crystallographic isotropic state $f(\boldsymbol{Q}) = 1$ delivers vanishing texture coefficients and a single crystal, i.e., $f(\boldsymbol{Q}) = \delta(\hat{\boldsymbol{Q}}, \boldsymbol{Q})$, possesses texture coefficients with maximum norm. It is shortly remarked that, in some literature references, the factor $(1 + 2\alpha)$ is sometimes incorporated either in the reference tensors $\mathbb{H}'_{\langle\alpha\rangle\beta}$ or in the texture coefficients $\mathbb{V}'_{\langle\alpha\rangle\beta}$, changing their norms. In this work the factor $(1 + 2\alpha)$ is singled out in order to work with normalized reference tensors and normalized texture coefficients.

Symmetry groups. Due to the properties of the CODF, see (2.27), the reference tensors have to reflect the microscopic symmetry of the material, i.e.,

$$\mathbb{H}'_{\langle\alpha\rangle\beta} = \boldsymbol{S} \star \mathbb{H}'_{\langle\alpha\rangle\beta} \quad \forall \boldsymbol{S} \in S_{\text{micro}} \,. \tag{2.45}$$

These constraints can decrease the number of basis harmonic tensors n'_α needed for the CODF. For example, for triclinic single crystal behavior, $S_{\text{micro}} = \{\boldsymbol{I}\}$ and $n'_4 = 1 + 2 \times 4 = 9$ hold, meaning that 9 linearly independent fourth-order reference tensors are required for the CODF. But if, e.g., cubic single crystal behavior is considered, then $n'_4 = 1 \leq 9$ holds and only one cubic fourth-order reference tensor exists.

Due to (2.28), the texture coefficients have to reflect the macroscopic sample symmetry

$$\mathbb{V}'_{\langle\alpha\rangle\beta} = \boldsymbol{S} \star \mathbb{V}'_{\langle\alpha\rangle\beta} \quad \forall \boldsymbol{S} \in S_{\text{macro}} \,. \tag{2.46}$$

These constraints allow reducing the number of free components of the texture coefficients. For example, although a microscopically cubic material might be considered, such that only one fourth-order cubic reference tensor, say $\mathbb{H}'_{\langle 4\rangle 1}$, is needed, the corresponding fourth-order texture coefficient, say $\mathbb{V}'_{\langle 4\rangle 1}$, might be arbitrarily triclinic. For this case $\mathbb{V}'_{\langle 4\rangle 1}$ would have 9 free components. If it is known that the sample is, e.g., orthotropic, then the number of free components of the fourth-order texture coefficient $\mathbb{V}'_{\langle 4\rangle 1}$ is reduced to 3, see, e.g., Appendix A.

Set of all possible texture coefficients. Due to the properties of the general CODF, the definition of the texture coefficients Eq. (2.44) can be interpreted as a convex combination of single crystal states. The set $\mathcal{V}$ of all possible texture coefficients is, therefore, the convex hull of all these single crystal states. In order to show this, following Lobos Fernández and Böhlke (2018), the notation

$$[[\mathbb{H}'_{\langle\alpha\rangle\beta}]] = (\, \mathbb{H}'_{\langle 1\rangle 1}, \ldots, \mathbb{H}'_{\langle 2\rangle 1}, \ldots\,) \tag{2.47}$$

is introduced to address all normalized reference tensors as an infinite-dimensional super vector. We address the rotation of such an infinite-dimensional super vector by

$$\boldsymbol{Q} \star [[\mathbb{H}'_{\langle\alpha\rangle\beta}]] = [[\boldsymbol{Q} \star \mathbb{H}'_{\langle\alpha\rangle\beta}]] = [[\mathbb{F}'_{\langle\alpha\rangle\beta}(\boldsymbol{Q})]] \,. \tag{2.48}$$

The set of all possible texture coefficients $\mathcal{V}$ is defined as

$$\mathcal{V} = \left\{ [[\mathbb{V}'_{\langle\alpha\rangle\beta}]] \mid \mathbb{V}'_{\langle\alpha\rangle\beta} = \langle f, \mathbb{F}'_{\langle\alpha\rangle\beta} \rangle \right\} \,. \tag{2.49}$$

with arbitrary CODF $f(\boldsymbol{Q})$. The set $\mathcal{V}$ corresponds to the limited microstructure hull of Adams et al. (2013), since $\mathcal{V}$ considers only one-point microstructure-information in Fourier space. The microstructure hull of Adams et al. (2013) is defined as the set of all possible microstructures that could exist in the material of interest, such that it contains the set $\mathcal{V}$. The set

$$\hat{\mathcal{V}} = \left\{ [[\mathbb{V}'_{\langle\alpha\rangle\beta}]] \mid \mathbb{V}'_{\langle\alpha\rangle\beta} = \sum_{\gamma=1}^{\infty} w_\gamma \mathbb{F}'_{\langle\alpha\rangle\beta}(\hat{\boldsymbol{Q}}_\gamma), w_\gamma \geq 0 \forall \gamma, \sum_{\gamma=1}^{\infty} w_\gamma = 1 \right\} \tag{2.50}$$

represents all possible convex combinations of all possible single crystal states. It is obvious that $\hat{\mathcal{V}} \subset \mathcal{V}$ holds since convex combinations of Dirac distributions are a special case of the general CODF. Consider now the infinite-dimensional vector function

$$I_{\hat{\mathcal{V}}}([[\mathbb{V}'_{\langle\alpha\rangle\beta}]]) = \begin{cases} 0 & [[\mathbb{V}'_{\langle\alpha\rangle\beta}]] \in \hat{\mathcal{V}} \\ +\infty & \text{else} \end{cases} . \tag{2.51}$$

This function is known in literature and is called the characteristic function of the considered set (in this case, of $\hat{\mathcal{V}}$). The characteristic function is a convex function. Jensen's inequality states that for a random variable X, the value of its expectation value $\mathbb{E}[X]$ mapped through a convex function ϕ fulfills $\phi(\mathbb{E}[X]) \leq \mathbb{E}[\phi(X)]$. Jensen's inequality can be applied to the convex function $I_{\hat{\mathcal{V}}}$, such that

$$I_{\hat{\mathcal{V}}}\left(\int_{SO(3)} f(\boldsymbol{Q})[[\mathbb{F}'_{\langle\alpha\rangle\beta}(\boldsymbol{Q})]]\mathrm{d}Q\right) \leq \int_{SO(3)} f(\boldsymbol{Q})\underbrace{I_{\hat{\mathcal{V}}}\left([[\mathbb{F}'_{\langle\alpha\rangle\beta}(\boldsymbol{Q})]]\right)}_{=0}\mathrm{d}Q \tag{2.52}$$

holds. The right-hand side of the inequality vanishes since every single crystal state is contained in $\hat{\mathcal{V}}$. It follows that the left-hand side has to vanish for all $f(\boldsymbol{Q})$, which means that all texture coefficients contained in $\mathcal{V}$ are also contained in $\hat{\mathcal{V}}$, i.e., $\mathcal{V} \subset \hat{\mathcal{V}}$. Finally, $\mathcal{V} = \hat{\mathcal{V}}$ holds.

It should be remarked that $\mathcal{V}$ is a subset of the set

$$\mathcal{N} = \{[[\mathbb{V}'_{\langle\alpha\rangle\beta}]] \mid \|\mathbb{V}'_{\langle\alpha\rangle\beta}\| \leq 1 \forall\alpha,\beta\} . \tag{2.53}$$

The set $\mathcal{N}$ describes the crystallographic anisotropy strength of the CODF, since single crystals are found on the boundary while polycrystals with a smooth texture deliver texture coefficients towards the origin.

Analytic expressions for $\mathcal{N}$ are trivial. Analytic expressions for $\mathcal{V}$ are difficult to find, especially since these depend on the microscopic and macroscopic symmetry groups. Even if only texture coefficients up to tensor order α are considered, with corresponding sets to be denoted as $\mathcal{V}_\alpha$, $\hat{\mathcal{V}}_\alpha$ and $\mathcal{N}_\alpha$, analytic expressions for the parametrization of the respective sets are still cumbersome. But since $\mathcal{V} = \hat{\mathcal{V}}$ holds, subsets $\check{\mathcal{V}}^{\mathrm{sym}}$ of $\hat{\mathcal{V}}$ defined as

$$\begin{aligned}\check{\mathcal{V}}^{\mathrm{sym}} = \Big\{[[\mathbb{V}'_{\langle\alpha\rangle\beta}]] \mid \quad \mathbb{V}'_{\langle\alpha\rangle\beta} = \sum_{\gamma=1}^{n} w_\gamma \mathrm{gs}^{\mathrm{sym}}(\mathbb{F}'_{\langle\alpha\rangle\beta}(\hat{\boldsymbol{Q}}_\gamma))\,, \\ w_\gamma \geq 0 \forall\gamma, \sum_{\gamma=1}^{n} w_\gamma = 1\Big\}\end{aligned} \qquad (2.54)$$

and corresponding projections $\check{\mathcal{V}}_\alpha^{\mathrm{sym}}$ can be constructed by consideration of a finite number n of single crystal orientations (generated in this work by Q_n, see Eq. (2.19)) and group symmetrization (see Eq. (2.6)) if a special symmetry group S^{sym} (see, e.g., Eq. (2.17)) is to be examined. Due to the construction based on single crystal orientations $\hat{\boldsymbol{Q}}_\gamma$, the original corresponding CODFs are based on Dirac distributions. Some group symmetrizations will maximize the norm in the respective Fourier directions. This does not have to be the case for every group symmetrization of a single crystal orientation. Some of these group symmetrizations approximate isotropic single crystal distributions up to a respective tensor order and deliver, therefore, texture coefficients near or even at the origin. But, since the corners of the set $\hat{\mathcal{V}}$ are single crystal orientations, it is clear that the boundary of every symmetry group dependent subset of $\hat{\mathcal{V}}$ is built by convex combinations of the corners of $\hat{\mathcal{V}}$. The convex set $\check{\mathcal{V}}^{\mathrm{sym}}$ is a trackable approximation of the exact symmetry group subset of $\hat{\mathcal{V}}$.

An example for polycrystals of hexagonal materials being axially symmetric to the direction $\boldsymbol{b}_3$ with aligned macroscopic hexagonal crystallographic symmetry (i.e., also axially symmetric to the direction $\boldsymbol{b}_3$) is presented based on S^{hex} (see Eq. (2.17)) for clarification. In this example only second- and fourth-order reference tensors will be investigated. Only one second-order and only one fourth-order hexagonal reference tensors exist. This implies that only one second-order $\mathbb{V}'_{\langle 2\rangle 1}$ and only one fourth-order $\mathbb{V}'_{\langle 4\rangle 1}$ texture coefficient exist. If the sample is to be hexagonal with respect to $\boldsymbol{b}_3$, then $\mathbb{V}'_{\langle 2\rangle 1}$ and $\mathbb{V}'_{\langle 4\rangle 1}$ are fully determined by their $(\mathbb{V}'_{\langle 2\rangle 1})_{11} = V'_{11}$ and $(\mathbb{V}'_{\langle 4\rangle 1})_{1111} = V'_{1111}$ components, respectively, see Appendix A. The set $\mathcal{V}_4^{\text{hex}}$ is known from literature as, see, e.g., Nomura et al. (1970), Müller and Böhlke (2016) and, more recerently, Lobos Fernández and Böhlke (2018),

$$\mathcal{V}_4^{\text{hex}} = \{(V'_{11}, V'_{1111}) \mid 10V'_{11}\left(\sqrt{6} - 21V'_{11}\right) + 12\sqrt{70}V'_{1111} + 7 \geq 0 \\ \wedge 5\sqrt{6}V'_{11} + 7 \geq 8\sqrt{70}V'_{1111}\} \,. \tag{2.55}$$

This set is depicted in yellow in Fig. 2.2.

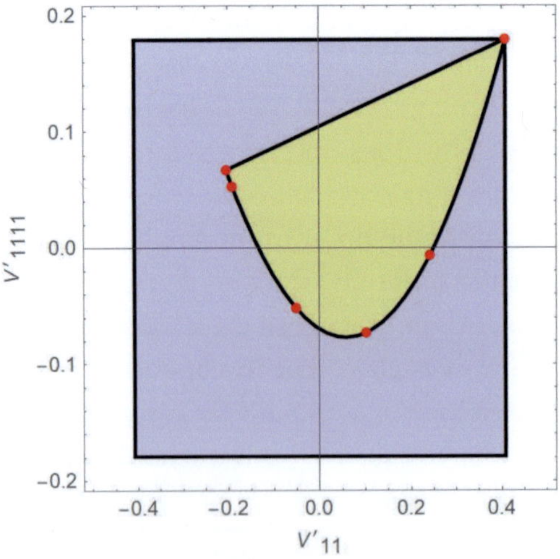

Figure 2.2: Sets $\mathcal{V}_4^{\text{hex}}$ and $\mathcal{N}_4^{\text{hex}}$ in yellow and blue, respectively; the convex hull of the red points delivers the set $\check{\mathcal{V}}_4^{\text{hex}}$ for this section

The corresponding set

$$\mathcal{N}_4^{\text{hex}} = \{(V'_{11}, V'_{1111}) \mid 6(V'_{11})^2 \leq 1 \wedge 280(V'_{1111})^2 \leq 9\} \tag{2.56}$$

is depicted in blue in Fig. 2.2. Finally, using S^{hex} (see Eq. (2.17)), group symmetrization (see Eq. (2.6)), and Q_n with $n = 4$ (see Eq. (2.19)), the set $\check{\mathcal{V}}_4^{\text{hex}}$ is generated as the convex hull of the red points depicted in Fig. 2.2. For this case, all red points of $\check{\mathcal{V}}_4^{\text{hex}}$ lie on the boundary of $\mathcal{V}_4^{\text{hex}}$.

The delineation of the set $\mathcal{V}$ is an important task not only for its mathematical meaning but also relevant for material models of polycrystalline and anisotropic media. Different models have been proposed in the literature to explore the evolution of the texture coefficients, see, e.g., Clément and Coulomb (1979), Morawiec et al. (1991), Böhlke and Bertram (2001b), Böhlke (2001), Böhlke (2006), Li et al. (2007), where the approach of Böhlke (2006) is assured to deliver texture coefficients in $\mathcal{V}$. Whatever model is proposed, it is crucial to investigate if the proposed model / differential equation for the evolution of the texture coefficients yields quantities at all times which stay in $\mathcal{V}$. To be clear, e.g., a material model might always deliver harmonic tensors in $\mathcal{N}$. But these tensors are not real texture coefficients if they are not in $\mathcal{V}$. If this is the case for a model, it would imply that the chosen material model delivers non-physical behavior, since there exists no CODF (and no real polycrystal or physical system) with texture coefficients outside of $\mathcal{V}$. This naturally also holds for models which might try to describe the evolution of texture eigenvalues in Λ. If an analytic description of $\mathcal{V}$ is not at hand, then the motivated subset $\check{\mathcal{V}}_\alpha^{\text{sym}}$ up to α tensor order offers a trackable option, where the single crystal orientations are to be chosen "properly". This could be achieved (1) with approximately equidistant distribution of a fixed but high number of orientations in $SO(3)$, see, e.g., Helming (1997) or Rosca et al. (2014), or (2) using a simple discretization as Q_n in Eq. (2.19), or (3) by consideration of iterations of increasing number single crystal orientations until the volume of the generated $\check{\mathcal{V}}_\alpha^{\mathrm{sym}}$ reaches a stationary numerical value under prescribed tolerances. The generation of $\check{\mathcal{V}}_\alpha^{\mathrm{sym}}$ is illustrative and instructive. It offers not only a trackable generating option but also trackable deductive perspective. If texture coefficients with degrees of freedom on the boundary of the respective $\mathcal{V}_\alpha$ are needed (e.g., due to optimization based on the texture coefficients), then single crystal CODFs can be tracked delivering these texture coefficients with maximum possible anisotropy. Alternatively, if the generated boundary points of $\check{\mathcal{V}}_\alpha^{\mathrm{sym}}$ cover the origin, as, e.g., in Fig. 2.2, then an infinite number of convex combinations and corresponding CODFs of the single crystal states can be generated to deliver vanishing texture coefficients up to tensor order α. For exact solutions for the special case of $\alpha = 4$ and polycrystals of cubic materials, see Bertram et al. (2000) and Böhlke and Bertram (2001a), and Gaffke et al. (2002) for polycrystals of general materials.

Alternative complex-valued Fourier expansion. The CODF may also be expanded in terms of complex-valued functions $T^l_{mn}(\boldsymbol{Q})$ as

$$f(\boldsymbol{Q}) = \sum_{l=0}^{\infty} \sum_{m,n=-l}^{l} C^l_{mn} T^l_{mn}(\boldsymbol{Q}) \,, \tag{2.57}$$

see, e.g., Roe (1965) and Bunge (1982). The complex-valued functions $T^l_{mn}(\boldsymbol{Q})$, with complex conjugate $T^{*l}_{mn}(\boldsymbol{Q})$, are usually chosen as to fulfill

$$\langle T^l_{mn}, T^{*\tilde{l}}_{\tilde{m}\tilde{n}} \rangle = \frac{\delta_{l\tilde{l}} \delta_{m\tilde{m}} \delta_{n\tilde{n}}}{(1+2l)} \,, \tag{2.58}$$

e.g., by using Wigner D-functions, see Wigner (1931). The Fourier coefficients C^l_{mn} are determined by

$$C^l_{mn} = (1+2l) \langle f, T^{*l}_{mn} \rangle \,. \tag{2.59}$$

The reader should note, that the Fourier expansion Eq. (2.57) has no tensor character. The TFE, see Eq. (2.43), has tensor character such that the tensorial texture coefficients $\mathbb{V}'_{\langle\alpha\rangle\beta}$ may be more attractive for the description of physical quantities in micromechanical models, instead of the coefficients C^l_{mn}. Nevertheless, the coefficients C^l_{mn} are vastly used in the field of texture analysis. For readers having access to the coefficients C^l_{mn} through experiments, the relation of these coefficients with the tensorial texture coefficients $\mathbb{V}'_{\langle\alpha\rangle\beta}$ may be useful in order to use the measured data in micromechanical models given in terms of the tensorial texture coefficients. Based on the properties of the respective function spaces spanned by the functions $\mathbb{F}'_{\langle\alpha\rangle\beta}(\boldsymbol{Q})$ and $T^l_{mn}(\boldsymbol{Q})$, it can be shown that the relations

$$
\begin{aligned}
C^l_{mn} &= (1+2l)\sum_{\beta=1}^{n'_l} \mathbb{V}'_{\langle l\rangle\beta} \cdot \langle \mathbb{F}'_{\langle l\rangle\beta}, T^{*l}_{mn}\rangle \, , \\
\mathbb{V}'_{\langle\alpha\rangle\beta} &= \sum_{m,n=-\alpha}^{\alpha} C^{\alpha}_{mn} \langle T^{\alpha}_{mn}, \mathbb{F}'_{\langle\alpha\rangle\beta}\rangle
\end{aligned}
\tag{2.60}
$$

hold, see Lobos Fernández and Böhlke (2018) for a detailed discussion and examples. The relation Eq. (2.60) is shortly motivated by consideration of the TFE for general triclinic materials based on the full space of harmonic tensors with orthonormal basis $\{\hat{\mathbb{H}}'_{\langle\alpha\rangle\beta}\}$ with $\beta = 1,\ldots,1+2\alpha$ and $\hat{\mathbb{H}}'_{\langle\alpha\rangle\beta} \cdot \hat{\mathbb{H}}'_{\langle\alpha\rangle\tilde{\beta}} = \delta_{\beta\tilde{\beta}}$. The corresponding functions $\hat{\mathbb{F}}'_{\langle\alpha\rangle\beta}(\boldsymbol{Q}) = \boldsymbol{Q} \star \hat{\mathbb{H}}'_{\langle\alpha\rangle\beta}$ are also harmonic tensors such that they may also be represented in terms of the basis $\{\hat{\mathbb{H}}'_{\langle\alpha\rangle\beta}\}$ for corresponding α, say, $\hat{\mathbb{F}}'_{\langle\alpha\rangle\beta}(\boldsymbol{Q}) = \sum_{\gamma=1}^{1+2\alpha} \hat{f}_{\alpha\beta\gamma}(\boldsymbol{Q})\hat{\mathbb{H}}'_{\langle\alpha\rangle\beta}$. The real-valued functions $\hat{f}_{\alpha\beta\gamma}(\boldsymbol{Q})$ form an orthogonal system with respect to $L^2(SO(3))$ and fulfill due to Eq. (2.42) $\langle \hat{f}_{\alpha\beta\gamma}, \hat{f}_{\tilde{\alpha}\tilde{\beta}\tilde{\gamma}}\rangle = \delta_{\alpha\tilde{\alpha}}\delta_{\beta\tilde{\beta}}\delta_{\gamma\tilde{\gamma}}/(1+2\alpha)$. The functions $\hat{f}_{\alpha\beta\gamma}(\boldsymbol{Q})$ span for given α the same $(1+2\alpha)^2$-dimensional real-valued function space M_α as the functions $T^l_{mn}(\boldsymbol{Q})$ for given $l=\alpha$, see Guidi et al. (1992). For alternative real-valued functions $T^l_{mn}(\boldsymbol{Q})$, see, e.g., Bunge (1982) or Schaeben and van den Boogaart (2003). Based on the orthogonality of the function spaces M_α and $M_{\tilde{\alpha}}$ for $\alpha \neq \tilde{\alpha}$, the relations Eq. (2.60) are obtained for triclinic materials and all special elasticity classes with corresponding symmetry groups, since the symmetry group dependent functions $\mathbb{F}'_{\langle\alpha\rangle\beta}$ are representable as linear combinations of $\hat{\mathbb{F}}'_{\langle\alpha\rangle\beta}$.

2.4 Harmonic decomposition

Introductory example: decomposition of a second-order tensor. The set of second-order tensors is shortly addressed as T_2. A second-order tensor $\tilde{\boldsymbol{D}} \in T_2$ can be decomposed as

$$\tilde{\boldsymbol{D}} = \boldsymbol{I}[\tilde{h}] + \boldsymbol{\epsilon}[\tilde{\boldsymbol{h}}] + \tilde{\boldsymbol{H}}' \tag{2.61}$$

where the scalar $\tilde{h}$, the vector $\tilde{\boldsymbol{h}}$ and the harmonic tensor $\tilde{\boldsymbol{H}}'$ contain the 9 degrees of freedom of $\tilde{\boldsymbol{D}}$. The isotropic part $\boldsymbol{I}[\tilde{h}] = \tilde{h}\boldsymbol{I}$, the skewed part $\boldsymbol{\epsilon}[\tilde{\boldsymbol{h}}]$ and the harmonic part $\tilde{\boldsymbol{H}}'$ of $\tilde{\boldsymbol{D}}$ belong to separate vector spaces which are invariant under the action of $SO(3)$. These vector spaces (i.e., the spaces of isotropic, T_2^I, skewed, T_2^s, and harmonic, T_2^h, second order tensors) can not be decomposed into smaller vector spaces which are also invariant under the action of $SO(3)$, i.e., these vector spaces are irreducible (under the action of $SO(3)$). Each of the tensors $\boldsymbol{I}[\tilde{h}], \boldsymbol{\epsilon}[\tilde{\boldsymbol{h}}], \tilde{\boldsymbol{H}}'$ is referred to as an irreducible tensor. The space T_2^I is one dimensional; it is induced by a scalar $\tilde{h}$ mapped into T_2^I by the isotropic tensor $\boldsymbol{I} \in T_2^I$. The space T_2^s is three dimensional; it is induced by the (axial) vector $\tilde{\boldsymbol{h}}$ mapped into T_2^s by the isotropic tensor $\boldsymbol{\epsilon} \in T_3^I$, T_3 being the set of third order tensors.

Definition. From the theory of group representations for $SO(3)$, see Schouten (1924) and Gel'fand et al. (1963), it is known that decompositions of a general r-th-order tensor $\tilde{\mathbb{D}}_{\langle r\rangle}$ in terms of tensors being irreducible under the action of elements of $SO(3)$ exist. The irreducible tensors might be given in terms of isotropic $(r+\alpha)$-th-order tensors $\mathbb{J}_{\langle r+\alpha\rangle\gamma}$ and harmonic α-th-order tensors $\mathbb{H}'_{\langle\alpha\rangle\gamma}$, such that the following decomposition

$$\tilde{\mathbb{D}}_{\langle r\rangle} = \sum_{\alpha=0}^{r}\sum_{\gamma=1}^{n_{r\alpha}} \mathbb{J}_{\langle r+\alpha\rangle\gamma}[\tilde{\mathbb{H}}'_{\langle\alpha\rangle\gamma}] = \mathrm{hd}(\tilde{\mathbb{H}}'_{\langle\alpha\rangle\gamma}) \tag{2.62}$$

exists. Each irreducible tensor $\mathbb{J}_{\langle r+\alpha\rangle\gamma}[\tilde{\mathbb{H}}'_{\langle\alpha\rangle\gamma}]$ from a corresponding irreducible tensor valued vector space is induced by a harmonic tensor $\tilde{\mathbb{H}}'_{\langle\alpha\rangle\gamma}$. This decomposition is known as the harmonic decomposition, see, e.g., Jerphagnon et al. (1978) or Auffray (2015). The harmonic tensors $\tilde{\mathbb{H}}'_{\langle\alpha\rangle\gamma}$ are referred to as harmonic parts. The harmonic parts for $\alpha = 0$ are scalars and represent, together with the corresponding isotropic tensors $\mathbb{J}_{\langle r\rangle\gamma}$, the isotropic part of the decomposed tensor. The harmonic parts for $\alpha \geq 1$ carry the anisotropy of the tensor and are referred to as anisotropic harmonic parts. The harmonic decomposition is not unique, i.e., a different set of suitable harmonic parts with corresponding isotropic tensors can also represent the original tensor, see, e.g., Auffray (2015). For algebraic approaches to such decompositions for arbitrary r-th-order tensors, see Spencer (1970), Zheng and Zou (2000) and Zou et al. (2001).

Due to the isotropy of the tensors $\mathbb{J}_{\langle r+\alpha\rangle\gamma}$, any orthogonal transformation applied to the original tensor is transferred to the harmonic parts

$$\boldsymbol{R} \star \tilde{\mathbb{D}}_{\langle r\rangle} = \sum_{\alpha=0}^{r}\sum_{\gamma=1}^{n_{r\alpha}} \mathbb{J}_{\langle r+\alpha\rangle\gamma}[\boldsymbol{R} \star \tilde{\mathbb{H}}'_{\langle\alpha\rangle\gamma}] \quad \forall \boldsymbol{R} \in Orth\,. \tag{2.63}$$

This implies that the symmetry group of $\tilde{\mathbb{D}}_{\langle r\rangle}$ is transferred to all anisotropic harmonic parts. These properties are naturally transferred to orientations $\boldsymbol{Q} \in SO(3) \subset Orth$ and corresponding symmetry groups $S \subset SO(3) \subset Orth$.

The harmonic parts $\tilde{\mathbb{H}}'_{\langle\alpha\rangle\gamma}$ carry the degrees of freedom and the symmetry group of the decomposed tensor $\tilde{\mathbb{D}}_{\langle r\rangle}$, i.e., they depend on $\tilde{\mathbb{D}}_{\langle r\rangle}$. The isotropic tensors $\mathbb{J}_{\langle r+\alpha\rangle\gamma}$ map linearly from the space of harmonic tensors and are constant tensors, i.e., independent of $\tilde{\mathbb{D}}_{\langle r\rangle}$.

The number $n_{r\alpha}$ in Eq. (2.62) gives the number of independent harmonic tensors of tensor order α for the harmonic decomposition of an r-th-order tensor. Depending on (1) index symmetries, (2) the symmetry group and (3) the specific components of the tensor $\tilde{\mathbb{D}}_{\langle r\rangle}$, different numbers $n_{r\alpha}$ of independent harmonic tensors exist. For example, for an arbitrary fourth-order tensor ($r = 4$), three scalars ($\alpha = 0$), six vectors ($\alpha = 1$), six second-order tensors ($\alpha = 2$), three third-order tensors ($\alpha = 3$) and one fourth-order tensor ($\alpha = 4 = r$) exist, which contain the total of $3 \times 1 + 6 \times 3 + 6 \times 5 + 3 \times 7 + 1 \times 9 = 81$ degrees of freedom of the general fourth-order tensor, see Schouten (1924). If the fourth-order tensor $\tilde{\mathbb{D}}_{\langle r\rangle}$ is minor symmetric in the left and right index pairs, then two scalars, one vector, three second-order tensors, one third-order tensor and one fourth-order tensor exist, which deliver the expected 36 degrees of freedom. If the minor symmetric fourth-order tensor $\tilde{\mathbb{D}}_{\langle r\rangle}$ possesses $SO(3)$ as its symmetric group (isotropic case), then only the scalars do not vanish. Finally, the components of this isotropic tensor can be chosen as to deliver equal scalar harmonic parts.

Representation through TFE. Consider the square integrable r-th-order tensor valued function $\mathbb{D}_{\langle r\rangle}(\boldsymbol{Q}) = \boldsymbol{Q} \star \tilde{\mathbb{D}}_{\langle r\rangle}$ and its TFE based on the functions $\mathbb{F}'_{\langle\alpha\rangle\beta}(\boldsymbol{Q})$, see Eq. (2.41), as for the CODF, see Eq. (2.43),

$$\mathbb{D}_{\langle r\rangle}(\boldsymbol{Q}) = \sum_{\alpha=0}^{\infty} \sum_{\beta=1}^{n'_\alpha} (1 + 2\alpha) \mathbb{V}_{\langle r+\alpha\rangle\beta}[\mathbb{F}'_{\langle\alpha\rangle\beta}(\boldsymbol{Q})] \,. \qquad (2.64)$$

The tensorial Fourier coefficients $\mathbb{V}_{\langle r+\alpha\rangle\beta}$ map linearly from the space of α-th-order harmonic tensors into the space of arbitrary r-th-order tensors. The tensorial Fourier coefficients $\mathbb{V}_{\langle r+\alpha\rangle\beta}$ are determined based on the orthogonality of the functions $\mathbb{F}'_{\langle\alpha\rangle\beta}$ as

$$\mathbb{V}_{\langle r+\alpha\rangle\beta} = \langle \mathbb{D}_{\langle r\rangle}, \mathbb{F}'_{\langle\alpha\rangle\beta}\rangle = \int\limits_{SO(3)} \mathbb{D}_{\langle r\rangle}(\boldsymbol{Q}) \otimes \mathbb{F}'_{\langle\alpha\rangle\beta}(\boldsymbol{Q})\mathrm{d}Q \,. \tag{2.65}$$

Now consider the harmonic decomposition of $\tilde{\mathbb{D}}_{\langle r\rangle}$. Equation (2.63) and a set of basis harmonic α-th-order tensors $\mathbb{B}'_{\langle\alpha\rangle\beta}$ with $\beta = 1, \ldots, n'_\alpha$ deliver

$$\begin{aligned} \mathbb{D}_{\langle r\rangle}(\boldsymbol{Q}) = \boldsymbol{Q} \star \tilde{\mathbb{D}}_{\langle r\rangle} &= \sum_{\alpha=0}^{r}\sum_{\gamma=1}^{n_{r\alpha}} \mathbb{J}_{\langle r+\alpha\rangle\gamma} \left[\sum_{\beta=1}^{n'_\alpha} \tilde{h}_{\alpha\gamma\beta}\boldsymbol{Q} \star \mathbb{B}'_{\langle\alpha\rangle\beta}\right] , \\ \tilde{\mathbb{H}}'_{\langle\alpha\rangle\gamma} &= \sum_{\beta=1}^{n'_\alpha} \tilde{h}_{\alpha\gamma\beta}\mathbb{B}'_{\langle\alpha\rangle\beta} \,. \end{aligned} \tag{2.66}$$

Rearranging the sums yields

$$\begin{aligned} \mathbb{D}_{\langle r\rangle}(\boldsymbol{Q}) = \boldsymbol{Q} \star \tilde{\mathbb{D}}_{\langle r\rangle} &= \sum_{\alpha=0}^{r}\sum_{\beta=1}^{n'_\alpha} \tilde{\mathbb{V}}_{\langle r+\alpha\rangle\beta}[\boldsymbol{Q} \star \mathbb{B}'_{\langle\alpha\rangle\beta}] , \\ \tilde{\mathbb{V}}_{\langle r+\alpha\rangle\beta} &= \sum_{\gamma=1}^{n_{r\alpha}} \tilde{h}_{\alpha\gamma\beta}\mathbb{J}_{\langle r+\alpha\rangle\gamma} \end{aligned} \tag{2.67}$$

Consideration of the symmetry group dependent reference tensors $\mathbb{H}'_{\langle\alpha\rangle\beta}$ for the basis harmonic tensors $\mathbb{B}'_{\langle\alpha\rangle\beta}$ delivers the perspective on Eq. (2.67) (compared to Eq. (2.64)) as the up to the tensor order r finite TFE of $\mathbb{D}_{\langle r\rangle}(\boldsymbol{Q}) = \boldsymbol{Q} \star \tilde{\mathbb{D}}_{\langle r\rangle}$ and the harmonic decomposition Eq. (2.62) as the finite TFE evaluated at $\boldsymbol{Q} = \boldsymbol{I}$. The corresponding isotropic tensorial Fourier coefficients $\mathbb{V}_{\langle r+\alpha\rangle\beta} = \tilde{\mathbb{V}}_{\langle r+\alpha\rangle\beta}$ are isotropic and material dependent. The functions $\boldsymbol{Q} \star \mathbb{B}'_{\langle\alpha\rangle\beta} = \boldsymbol{Q} \star \mathbb{H}'_{\langle\alpha\rangle\beta} = \mathbb{F}'_{\langle\alpha\rangle\beta}(\boldsymbol{Q})$ are material independent but symmetry group dependent.

Example: fourth-order tensor with minor symmetries. A fourth-order tensor $\tilde{\mathbb{D}}$ with minor symmetries, i.e., $\tilde{D}_{ijkl} = \tilde{D}_{jikl} = \tilde{D}_{ijlk}$, but not necessarily major symmetric can be decomposed as proposed in Spencer (1970) into its harmonic decomposition (see also Forte and Vianello (1997)). For the case of an additionally major symmetric tensor, see Forte and Vianello (1996). In this work, the harmonic decomposition of Forte and Vianello (1997) is presented following Lobos et al. (2017) and Lobos Fernández and Böhlke (2018) as follows

$$
\begin{aligned}
\tilde{\mathbb{D}} &= \mathrm{hd}(\tilde{h}_{I1}, \tilde{h}_{I2}, \tilde{\boldsymbol{h}}, \tilde{\boldsymbol{H}}_1', \tilde{\boldsymbol{H}}_2', \tilde{\boldsymbol{H}}_3', \tilde{\boldsymbol{H}}_{\langle 3\rangle}', \tilde{\mathbb{H}}') \\
&= \tilde{h}_{I1}\mathbb{P}_1 + \tilde{h}_{I2}\mathbb{P}_2 + \mathbb{J}_{\langle 5\rangle}[\tilde{\boldsymbol{h}}] + \mathbb{J}_{\langle 6\rangle 1}[\tilde{\boldsymbol{H}}_1'] + \mathbb{J}_{\langle 6\rangle 2}[\tilde{\boldsymbol{H}}_2'] \qquad (2.68) \\
&\quad + \mathbb{J}_{\langle 6\rangle 3}[\tilde{\boldsymbol{H}}_3'] + \mathbb{J}_{\langle 7\rangle}[\tilde{\boldsymbol{H}}_{\langle 3\rangle}'] + \tilde{\mathbb{H}}' \, .
\end{aligned}
$$

The scalars $\tilde{h}_{I1,2}$, the vector $\tilde{\boldsymbol{h}}$ and the fully symmetric and traceless tensors $\tilde{\boldsymbol{H}}_\beta'$, $\tilde{\boldsymbol{H}}_{\langle 3\rangle}'$ and $\tilde{\mathbb{H}}'$ reflect the $2 \times 1 + 3 + 3 \times 5 + 7 + 9 = 36$ free components of minor symmetric fourth-order tensors. The isotropic tensors $\mathbb{J}_{\langle\alpha\rangle\beta}$ are given explicitly in Lobos et al. (2017), see alternatively Appendix B. The scalars $\tilde{h}_{I1,2}$ represent the isotropic part of the decomposed tensor, while the rest describes the anisotropy and symmetry group of the tensor. A straightforward implementation in *Mathematica*® 11 for the computation of this harmonic decomposition is given together with examples in Lobos et al. (2017). If the considered tensor $\tilde{\mathbb{D}}$ is additionally major symmetric, i.e., $\tilde{D}_{ijkl} = \tilde{D}_{klij}$, then the harmonic parts $\tilde{\boldsymbol{h}}$, $\tilde{\boldsymbol{H}}_3'$ and $\tilde{\boldsymbol{H}}_{\langle 3\rangle}'$ vanish.

2.5 The orientation average

2.5.1 Orientation average of a minor symmetric fourth-order tensor

The orientation average $\langle f, \mathbb{D} \rangle$ based on a minor symmetric fourth-order single crystal tensor $\tilde{\mathbb{D}}$ is now investigated. Two main perspectives on the orientation average will be considered in Section 2.5.2 and Section 2.5.3. The first one is the consideration of the orientation average as the simple linear map of the single crystal behavior $\tilde{\mathbb{D}}$ as follows

$$\langle f, \mathbb{D} \rangle = \mathbb{Q}^f_{\langle 8 \rangle}[\tilde{\mathbb{D}}] \,, \quad \mathbb{Q}^f_{\langle 8 \rangle} = \int\limits_{SO(3)} f(\boldsymbol{Q}) \boldsymbol{Q}^{\star 4} \mathrm{d}Q \,, \tag{2.69}$$

with the Rayleigh power $\boldsymbol{Q}^{\star r}$, defined as $\boldsymbol{Q}^{\star r}[\mathbb{K}_{\langle r \rangle}] = \boldsymbol{Q} \star \mathbb{K}_{\langle r \rangle}$ and components $(\boldsymbol{Q}^{\star r})_{i_1 \dots i_r j_1 \dots j_r} = Q_{i_1 j_1} \dots Q_{i_r j_r}$. As a linear map, several decompositions of $\mathbb{Q}^f_{\langle 8 \rangle}$ might be of interest. From the point of view of orientation averaging, the polar decomposition $\mathbb{Q}^f_{\langle 8 \rangle} = \mathbb{R}^f_{\langle 8 \rangle} \mathbb{U}^f_{\langle 8 \rangle}$ with stretching major symmetric positive semi-definite part $\mathbb{U}^f_{\langle 8 \rangle}$ and unitary part $\mathbb{R}^f_{\langle 8 \rangle}$ (sometimes corresponding to a rotation) is of special interest. If the stretching part is positive definite, then the unitary part is unique. Two main special cases are important here. For a single crystal with orientation $\hat{\boldsymbol{Q}}$, i.e., $f(\boldsymbol{Q}) = \delta(\hat{\boldsymbol{Q}}, \boldsymbol{Q})$, $\mathbb{Q}^\delta_{\langle 8 \rangle} = \mathbb{R}^\delta_{\langle 8 \rangle} \mathbb{U}^\delta_{\langle 8 \rangle}$ with $\mathbb{U}^\delta_{\langle 8 \rangle} = \mathbb{I}_{\langle 8 \rangle}$ and $\mathbb{R}^\delta_{\langle 8 \rangle} = \hat{\boldsymbol{Q}}^{\star 4}$ clearly holds, $\mathbb{I}_{\langle 8 \rangle}$ being the full identity on fourth-order tensors which is positive definite. For the crystallographic isotropic case, i.e., $f(\boldsymbol{Q}) = 1$, the averaging tensor is given by $\mathbb{Q}^1_{\langle 8 \rangle} = \mathbb{R}^1_{\langle 8 \rangle} \mathbb{U}^1_{\langle 8 \rangle}$ with the identity on isotropic fourth-order tensors as stretching part $\mathbb{U}^1_{\langle 8 \rangle} = \mathbb{I}^I_{\langle 8 \rangle}$ and $\mathbb{R}^1_{\langle 8 \rangle} = \boldsymbol{Q}^{\star 4}$ with arbitrary $\boldsymbol{Q} \in SO(3)$. In this case, $\mathbb{U}^1_{\langle 8 \rangle}$ contracts the single crystal behavior to its isotropic part, mapping any anisotropic part to zero. This means that, with respect to general fourth-order tensors, the identity on isotropic tensors has vanishing eigenvalues and is, therefore, positive semi-definite. As a consequence, $\mathbb{R}^1_{\langle 8\rangle}$ is not unique. This first perspective on the orientation average will be considered in Section 2.5.2 for CCODFs, as investigated in Lobos et al. (2017).

The second perspective on the orientation average is the direct application of the general harmonic decomposition Eq. (2.62) and its special case Eq. (2.68) for the minor symmetric tensor $\tilde{\mathbb{D}}$. Due to linearity and the property Eq. (2.63), the general orientation average is transmitted to the harmonic parts. For the orientation average $\langle f, \mathbb{D}\rangle$, the isotropic part remains unchanged while the orientation average of the first- to fourth-order harmonic parts is to be examined. This second perspective will be taken into consideration in Section 2.5.3, as in Lobos Fernández and Böhlke (2018).

2.5.2 Orientation average based on CCODFs

The orientation average based on $\mathbb{Q}^g_{\langle 2r\rangle}$. In this section, we consider the orientation average $\langle g, \mathbb{D}_{\langle r\rangle}\rangle$ of a general r-th-order tensor $\tilde{\mathbb{D}}_{\langle r\rangle}$ based on a CCODF $g(\hat{\omega})$ centered around $\hat{\boldsymbol{Q}}$, such that the $2r$-th-order orientation averaging tensor

$$\mathbb{Q}^g_{\langle 2r\rangle} = \int\limits_{SO(3)} g(\hat{\omega}) \boldsymbol{Q}^{\star r} \mathrm{d}Q \tag{2.70}$$

is to be investigated. The tensor $\mathbb{Q}^g_{\langle 2r\rangle}$ has been investigated in Lobos et al. (2017) for $r \in \{1, 2, 3, 4\}$ with detailed derivation. The objective of this section is to extend these results for arbitrary tensor order r.

Based on the shift property Eq. (2.24) of the general CODF, the orientation average might be shifted with the central orientation $\hat{\boldsymbol{Q}}$ of the CCODF

$$\mathbb{Q}^g_{\langle 2r\rangle} = \hat{\boldsymbol{Q}}^{\star r} \int\limits_{SO(3)} g(\omega)\boldsymbol{Q}^{\star r}\mathrm{d}Q = \hat{\boldsymbol{Q}}^{\star r} \int\limits_{\omega=0}^{\pi} g(\omega) \int\limits_{S_2} \boldsymbol{Q}^{\star r}\mathrm{d}n \; s(\omega)\, \mathrm{d}\omega \,. \tag{2.71}$$

Now, consider the parametrization of the orientation $\boldsymbol{Q}$ in terms of dyadic powers $\boldsymbol{N}_{0,1,2}$ of the rotation axis $\boldsymbol{n}$. The integral over S_2 eliminates any odd dyadic powers of $\boldsymbol{n}$. This holds due to the isotropic averaging over S_2 such that for each $\boldsymbol{n}$ its counterpart $-\boldsymbol{n}$ is weighted equally for all odd dyadic powers of $\boldsymbol{n}$ which then cancel each other out. Only even dyadic powers of $\boldsymbol{n}$ survive and the resulting integrals over S_2 are isotropic tensors. This is shown directly with the property Eq. (2.24) and $f(\boldsymbol{Q}) = 1$. Now, let d^I_{2r} denote the dimension of the space of $2r$-th-order isotropic tensors and $\mathbb{B}^I_{\langle 2r\rangle\beta}$ isotropic basis tensors. The isotropic tensor $\int_{S_2} \boldsymbol{Q}^{\star n}\mathrm{d}n$ must be representable by a linear combination of the isotropic basis tensors. The ω dependent coefficients $c_\beta(\omega)$ of the linear combination can be computed with standard projections and solving the corresponding linear system

$$\begin{aligned} \int\limits_{S_2} \boldsymbol{Q}^{\star r}\mathrm{d}n &= \sum_{\beta=1}^{d^I_{2r}} c_\beta(\omega)\mathbb{B}^I_{\langle 2r\rangle\beta} \,, \\ \int\limits_{S_2} \boldsymbol{Q}^{\star r} \cdot \mathbb{B}^I_{\langle 2r\rangle\alpha}\mathrm{d}n &= \sum_{\beta=1}^{d^I_{2r}} (\mathbb{B}^I_{\langle 2r\rangle\alpha} \cdot \mathbb{B}^I_{\langle 2r\rangle\beta})c_\beta(\omega) \,. \end{aligned} \tag{2.72}$$

The scalars $\boldsymbol{Q}^{\star r} \cdot \mathbb{B}^I_{\langle 2r\rangle\alpha}$ can be examined more clearly with the projection

$$p(\boldsymbol{n},\omega) = \boldsymbol{Q}^{\otimes r} \cdot \mathbb{B}_{\langle 2r\rangle} = (\boldsymbol{N}_0 + \boldsymbol{N}_1 + \boldsymbol{N}_2)^{\otimes r} \cdot \mathbb{B}_{\langle 2r\rangle} \tag{2.73}$$

with an arbitrary isotropic $2r$-th-order isotropic tensor $\mathbb{B}_{\langle 2r\rangle}$. This is possible since the Rayleigh power $\boldsymbol{Q}^{\star r}$ can be transformed into the dyadic power $\boldsymbol{Q}^{\otimes r}$ with a corresponding index permutation, which can be transferred through the scalar product to the isotropic tensor. Consider, e.g., $r=2$, $p_2=\boldsymbol{Q}^{\star 2}\cdot\mathbb{B}^I$ and an isotropic tensor $\mathbb{B}^I$. Short computation shows $p_2=\sum_{i,j,k,l=1}^{3} Q_{ik}Q_{jl}B^I_{ijkl}=\sum_{i,j,k,l=1}^{3} Q_{ij}Q_{kl}B_{ijkl}$ defining $\mathbb{B}$ by $B_{ijkl}=B^I_{ikjl}$, such that $p_2=\boldsymbol{Q}^{\star 2}\cdot\mathbb{B}^I=\boldsymbol{Q}^{\otimes 2}\cdot\mathbb{B}$ holds. Any index permutation / transposition of an isotropic tensor delivers just another isotropic tensor. The scalar p is an isotropic function of the rotation axis $\boldsymbol{n}$ of $\boldsymbol{Q}$, where $p(\boldsymbol{n},\omega)=p(\boldsymbol{R}\boldsymbol{n},\omega)\forall\boldsymbol{R}\in Orth$ is shown by transferring $\boldsymbol{R}$ to the isotropic tensor $\mathbb{B}_{\langle 2r\rangle}$. In order to see this, consider first $\boldsymbol{Q}(\boldsymbol{R}\boldsymbol{n},\omega)$, which can be reformulated as $\boldsymbol{Q}(\boldsymbol{R}\boldsymbol{n},\omega)=\boldsymbol{R}\star(\boldsymbol{Q}(\boldsymbol{n},\omega))$. Then, based on $(\boldsymbol{R}\star\boldsymbol{A})^{\otimes r}=\boldsymbol{R}\star(\boldsymbol{A}^{\otimes r})$ and $(\boldsymbol{R}\star\mathbb{A}_{\langle r\rangle})\cdot\hat{\mathbb{A}}_{\langle r\rangle}=\mathbb{A}_{\langle r\rangle}\cdot(\boldsymbol{R}^{\mathsf{T}}\star\hat{\mathbb{A}}_{\langle r\rangle})$, it follows, e.g., $\boldsymbol{Q}(\boldsymbol{R}\boldsymbol{n},\omega)^{\otimes 2}\cdot\mathbb{B}=\boldsymbol{Q}(\boldsymbol{n},\omega)^{\otimes 2}\cdot(\boldsymbol{R}^{\mathsf{T}}\star\mathbb{B})=\boldsymbol{Q}(\boldsymbol{n},\omega)^{\otimes 2}\cdot\mathbb{B}$ for all orthogonal $\boldsymbol{R}$ due to the isotropy of $\mathbb{B}$. Analogous reasoning implies the isotropy of p with respect to $\boldsymbol{n}$, such that p does not depend on $\boldsymbol{n}$. Further, the scalar p is an even function of the rotation angle ω, i.e., $p(\boldsymbol{n},\omega)=p(\boldsymbol{n},-\omega)$ holds. This is immediately shown by considering $p(\boldsymbol{n},-\omega)=p(\boldsymbol{R}\boldsymbol{n},\omega)$ with $\boldsymbol{R}=-\boldsymbol{I}\in Orth$ and by using the isotropy of p with respect to $\boldsymbol{n}$. Now, with finite powers of $\sin(\omega)$ and $\cos(\omega)$ up to r for $\boldsymbol{Q}^{\star r}$ (which, based on Eulers formula, can be expressed as linear combinations of $\sin(k\omega)$ and $\cos(k\omega)$ for $k\le r$), the scalar p (which is an even function of ω) can be represented by a finite series up to r of the functions $\mu_\alpha(\omega)$, see Eq. (2.34). This implies that every projection $\boldsymbol{Q}^{\star n}\cdot\mathbb{B}^I_{\langle 2r\rangle\alpha}$ can be represented as a linear combination of the functions $\mu_\alpha(\omega)$ up to r.

Due to the linear relations defined through Eq. (2.72), the coefficients $c_\beta(\omega)$ can be expressed as a linear combination of the functions $\mu_\alpha(\omega)$ up to r, say, $c_\beta(\omega) = \sum_{\alpha=1}^{r} c_{\beta\alpha} \int_{\mathcal{S}} dn\mu_\alpha(\omega)$. Therefore, replacing $\sum_{\beta=1}^{d_{2r}^I} c_{\beta\alpha}\mathbb{B}^I_{\langle 2r\rangle\beta} = \hat{\mathbb{B}}^I_{\langle 2r\rangle\alpha}$, the alternative representation

$$\int\limits_{S_2} \boldsymbol{Q}^{\star r} \mathrm{d}n = \sum_{\alpha=0}^{r} \int\limits_{S_2} \mathrm{d}n\, \mu_\alpha(\omega)\, \hat{\mathbb{B}}^I_{\langle 2r\rangle\alpha} \tag{2.74}$$

is obtained. It should be noted that $r \leq d_{2r}^I$ holds and, therefore, the representation Eq. (2.74) delivers a special decomposition of the general $2r$-th-order isotropic tensor. The functions $\mu_\alpha(\omega)$ build an orthogonal system, see Eq. (2.35). Multiplying Eq. (2.74) with $s(\omega)\mu_{\tilde{\alpha}}(\omega)$, see Eq. (2.12), and integrating over ω, see Eq. (2.16), allows to isolate the tensor $\hat{\mathbb{B}}^I_{\langle 2r\rangle\tilde{\alpha}}$. The isotropic tensors $\hat{\mathbb{B}}^I_{\langle 2r\rangle\alpha}$ are, therefore, identified as

$$\hat{\mathbb{B}}^I_{\langle 2r\rangle\alpha} = (1+2\alpha)^2 \int\limits_{SO(3)} \mu_\alpha(\omega)\boldsymbol{Q}^{\star r} \mathrm{d}Q\,, \tag{2.75}$$

with $\hat{\mathbb{B}}^I_{\langle 2r\rangle 0}$ being clearly the identity on isotropic r-th-order tensors $\mathbb{I}^I_{\langle 2r\rangle}$, since $\hat{\mathbb{B}}^I_{\langle 2r\rangle 0}$ corresponds to the orientation average with $1 = \mu_0(\omega)$. It should be noted that for $\alpha > r$, $\hat{\mathbb{B}}^I_{\langle 2r\rangle\alpha} = \mathbb{O}_{\langle 2r\rangle}$ holds, meaning that the finite series given Eq. (2.74) could be extended to infinity. The identity $\mathbb{I}^I_{\langle 2r\rangle}$ maps an arbitrary $\tilde{\mathbb{D}}_{\langle r\rangle}$ into its isotropic part. Based on this property, it can be shown that the scalar $\boldsymbol{Q}^{\star r} \cdot \mathbb{I}^I_{\langle 2r\rangle}$ is a constant, i.e., $\boldsymbol{Q}^{\star r} \cdot \mathbb{I}^I_{\langle 2r\rangle}$ does not depend on ω, and is, therefore, proportional to $\mu_0(\omega) = 1$, see Eq. (2.34). This implies that all $\hat{\mathbb{B}}^I_{\langle 2r\rangle\alpha}$, $\alpha \geq 1$, are orthogonal to $\hat{\mathbb{B}}^I_{\langle 2r\rangle 0} = \mathbb{I}^I_{\langle 2r\rangle}$. Further, by representing the Dirac distribution as $\delta(\omega) = \sum_{\alpha=0}^{\infty}(1+2\alpha)^2\mu_\alpha(\omega)$, $\sum_{\alpha=0}^{r} \hat{\mathbb{B}}^I_{\langle 2r\rangle\alpha} = \sum_{\alpha=0}^{\infty} \hat{\mathbb{B}}^I_{\langle 2r\rangle\alpha} = \mathbb{I}_{\langle 2r\rangle}$ holds, $\mathbb{I}_{\langle 2r\rangle}$ being the identity on general r-th-order tensors. This allows to compute the last tensor $\hat{\mathbb{B}}^I_{\langle 2r\rangle r}$ for fixed r based on the previous tensors.

As a result, the general averaging tensor $\mathbb{Q}^g_{\langle 2r\rangle}$ can be represented in terms of the texture eigenvalues λ_α of the general CCODF as follows

$$
\begin{aligned}
\mathbb{Q}^g_{\langle 2r\rangle} &= \hat{\boldsymbol{Q}}^{\star r}\left(\sum_{\alpha=0}^{r}\lambda_\alpha\hat{\mathbb{B}}^I_{\langle 2r\rangle\alpha}\right) \\
&= \hat{\boldsymbol{Q}}^{\star r}\left(\mathbb{I}^I_{\langle 2r\rangle}+\sum_{\alpha=1}^{r-1}\lambda_\alpha\hat{\mathbb{B}}^I_{\langle 2r\rangle\alpha}+\lambda_r\hat{\mathbb{B}}^I_{\langle 2r\rangle r}\right) .
\end{aligned}
\tag{2.76}
$$

This result generalizes the results of Lobos et al. (2017) to arbitrary tensor order and might be applied to arbitrarily non-symmetric r-th-order tensors. The limits for crystallographic isotropy ($\lambda_\alpha \to 0, \alpha \geq 1$) and for the single crystal ($\lambda_\alpha \to 1$) are visible in the obtained representation.

It should be noted that due to the linear map over isotropic tensors, if a general r-th-order tensor $\tilde{\mathbb{D}}_{\langle r\rangle}$ is mapped over $\mathbb{Q}^g_{\langle 2r\rangle}$, then the symmetry group of $\tilde{\mathbb{D}}_{\langle r\rangle}$ is automatically transferred to $\mathbb{Q}^g_{\langle 2r\rangle}[\tilde{\mathbb{D}}_{\langle r\rangle}]$, only reoriented by the central orientation $\hat{\boldsymbol{Q}}$. This means, that, e.g., if the single crystal property $\tilde{\mathbb{D}}_{\langle r\rangle}$ is monoclinic, then $\mathbb{Q}^g_{\langle 2r\rangle}[\tilde{\mathbb{D}}_{\langle r\rangle}]$ is also monoclinic, only with respect to the reference crystallographic axes rotated with $\hat{\boldsymbol{Q}}$.

There are cases for which the isotropic tensors $\hat{\mathbb{B}}^I_{\langle 2r\rangle\alpha}$ posses projector properties, i.e., $\hat{\mathbb{B}}^I_{\langle 2r\rangle\alpha}\hat{\mathbb{B}}^I_{\langle 2r\rangle\beta} = \mathbb{O}_{\langle 2r\rangle}\forall\alpha\neq\beta$, $\hat{\mathbb{B}}^I_{\langle 2r\rangle\alpha}\hat{\mathbb{B}}^I_{\langle 2r\rangle\alpha} = \hat{\mathbb{B}}^I_{\langle 2r\rangle\alpha}$ for given α. Due to the orthogonality of $\hat{\mathbb{B}}^I_{\langle 2r\rangle\alpha}, \alpha \geq 1$, with respect to $\mathbb{I}^I_{\langle 2r\rangle}$, the tensors $\hat{\mathbb{B}}^I_{\langle 2r\rangle\alpha}, \alpha \geq 1$ must map into anisotropic spaces of r-th-order tensors. For these cases, the linear combination of the isotropic tensors $\hat{\mathbb{B}}^I_{\langle 2r\rangle\alpha}$ delivers a special projector decomposition of the general isotropic $2r$-th-order tensor and the averaging tensor $\mathbb{Q}^g_{\langle 2r\rangle}$ might be decomposed into its polar decomposition $\mathbb{Q}^g_{\langle 2r\rangle} = \mathbb{R}^g_{\langle 2r\rangle}\mathbb{U}^g_{\langle 2r\rangle}$ with

$$
\mathbb{R}^g_{\langle 2r\rangle} = \hat{\boldsymbol{Q}}^{\star r}\left(\sum_{\alpha=0}^{r}\operatorname{sgn}(\lambda_\alpha)\hat{\mathbb{B}}^I_{\langle 2r\rangle\alpha}\right) , \quad \mathbb{U}^g_{\langle 2r\rangle} = \sum_{\alpha=0}^{r}|\lambda_\alpha|\hat{\mathbb{B}}^I_{\langle 2r\rangle\alpha} . \tag{2.77}
$$

Now, the linear map of an arbitrary r-th-order tensor $\tilde{\mathbb{D}}_{\langle r\rangle}$ over $\mathbb{Q}^g_{\langle 2r\rangle}$ for these cases can be discussed. The tensor $\tilde{\mathbb{D}}_{\langle r\rangle}$ is first decomposed into orthogonal parts by the tensors $\hat{\mathbb{B}}^I_{\langle 2r\rangle\alpha}$. These orthogonal parts of $\tilde{\mathbb{D}}_{\langle r\rangle}$ might be contracted concerning their norms by the absolute value of the anisotropic texture eigenvalues $|\lambda_\alpha| \leq 1$. But this contraction is applied only the the anisotropic parts of the mapped tensor $\tilde{\mathbb{D}}_{\langle r\rangle}$, since $\lambda_0 = 1$ holds. The tensor $\mathbb{U}^g_{\langle 2r\rangle}$ is, therefore, considered as a contraction map. The map over $\mathbb{R}^g_{\langle 2r\rangle}$ can be interpreted as the application of an eventual reflection of different anisotropic parts for $\mathrm{sgn}(\lambda_\alpha) = -1, \alpha \geq 1$, and a final reorientation with the central orientation $\hat{\boldsymbol{Q}}$. These results are not only in agreement with the results obtained in Lobos et al. (2017) for $\mathbb{Q}^g_{\langle 2r\rangle}, r = 1, 2, 3$, and 4, but hold for arbitrary tensor-order.

Cases with $\mathrm{sgn}(\lambda_\alpha) = -1, \alpha \geq 1$ might be interesting for tensorial physical quantities $\tilde{\mathbb{D}}_{\langle r\rangle}$ needed for certain applications but where an inversion of the anisotropic properties directions may be beneficial, see Lobos et al. (2017). Not all anisotropic property directions can be inverted by a single CCODF, since there exists no CCODF delivering only negative anisotropic texture eigenvalues. This is proven already by contemplation of the functions $\mu_\alpha(\omega), \alpha \in \{1, 2, 3, 4\}$, in Fig. 2.1a and consideration of the artificial CCODF $g^\delta(\omega) = \delta(\omega - \omega_0)/(4\pi s(\omega_0))$. There exists no ω_0 such that $\mu_\alpha(\omega_0) < 0 \; \forall \alpha \in \{1, 2, 3, 4\}$. This implies, due to convexity, that no convex combination, i.e., no $g(\omega)$, exists such that $\lambda_\alpha < 0 \; \forall \alpha \in \{1, 2, 3, 4\}$. This naturally rules out the general case for all texture eigenvalues. Nevertheless, certain cases, e.g., $\lambda_\alpha < 0 \; \forall \alpha \in \{1, 2\}$ or $\forall \alpha \in \{2, 4\}$, as visible in Fig. 2.1b, are possible, and more importantly, *independently of the symmetry group of the single crystal behavior.* This proves the existence of CCODFs (and therefore of general CODFs) being able to invert the direction of certain anisotropic properties independently of the kind of anisotropy itself.

For clarification, consider the case $r = 2$. With $\hat{\mathbb{B}}^I_{\langle 4\rangle 0} = \mathbb{P}_1$, $\hat{\mathbb{B}}^I_{\langle 4\rangle 1} = \mathbb{P}_3$ and $\hat{\mathbb{B}}^I_{\langle 4\rangle 2} = \mathbb{P}_2$, the averaging tensor for this case is given by

$$\mathbb{Q}^g_{\langle 4\rangle} = \hat{\boldsymbol{Q}}^{\star 2}(\mathbb{P}_1 + \lambda_2\mathbb{P}_2 + \lambda_1\mathbb{P}_3)\,, \tag{2.78}$$

see Lobos et al. (2017). An arbitrary second-order tensor $\tilde{\boldsymbol{D}}$ (corresponding to a single crystal material behavior) is decomposed by the projectors $\mathbb{P}_i$ into its isotropic, symmetric traceless and skewed part. The symmetry group of $\tilde{\boldsymbol{D}}$ is transferred to $\mathbb{Q}^g_{\langle 4\rangle}[\tilde{\boldsymbol{D}}]$. Only the anisotropic parts of $\tilde{\boldsymbol{D}}$ might be contracted with respect to their norm. An inversion of the anisotropic directions is always possible, i.e., $\mathrm{sgn}(\lambda_\alpha) = -1\ \forall\alpha \in \{1,2\}$, *independently of the anisotropic symmetry group of* $\tilde{\boldsymbol{D}}$. This is proven with $g^\delta(\omega) = \delta(\omega - \omega_0)/(4\pi s(\omega_0))$ and any ω_0 yielding $\mu_\alpha(\omega_0) < 0\forall\alpha \in \{1,2\}$, see Fig. 2.1a.

The isotropic tensors $\hat{\mathbb{B}}^I_{\langle 2r\rangle\alpha}$, see (2.75), for $r \in \{1,2,3,4\}$ have been determined computationally and are given explicitly in Appendix B. These tensors might be used for a rapid computation of the respective orientation averages of arbitrary r-th-order tensors. For an extension of these computations, algorithms implemented in *Mathematica*® 11 are also given in Appendix B in order to generate bases of isotropic tensors of arbitrary order which can then be used for the determination of the tensors $\hat{\mathbb{B}}^I_{\langle 2r\rangle\alpha}$ for $r \geq 5$.

Average of a minor symmetric fourth-order tensor. We consider now the orientation average $\langle g, \mathbb{D}\rangle$ for a minor symmetric fourth-order tensor $\tilde{\mathbb{D}}$. Either, the tensor $\mathbb{Q}^g_{\langle 8\rangle}$ can be considered in its full form, or, using the harmonic decomposition of $\tilde{\mathbb{D}}$, solely the corresponding orientation averages of the first-order to fourth-order harmonic parts $\tilde{\boldsymbol{h}}$, $\tilde{\boldsymbol{H}}'_\beta$, $\tilde{\boldsymbol{H}}'_{\langle 3\rangle}$ and $\tilde{\mathbb{H}}'$ are to be examined for a CCODF. In this work, the complete tensor $\mathbb{Q}^g_{\langle 8\rangle}$ is determined with the isotropic tensors $\hat{\mathbb{B}}^I_{\langle 8\rangle\alpha}$ given in Appendix B.

Lengthy but simple computations with $\mathbb{Q}^g_{\langle 8\rangle}$ deliver then

$$\langle g, \mathbb{D}\rangle = \hat{\boldsymbol{Q}} \star \mathrm{hd}(\tilde{h}_{I1}, \tilde{h}_{I2}, \lambda_1 \tilde{\boldsymbol{h}}, \lambda_2 \tilde{\boldsymbol{H}}'_1, \lambda_2 \tilde{\boldsymbol{H}}'_2, \lambda_2 \tilde{\boldsymbol{H}}'_3, \lambda_3 \tilde{\boldsymbol{H}}'_{\langle 3\rangle}, \lambda_4 \tilde{\mathbb{H}}') \ , \tag{2.79}$$

see also Lobos et al. (2017), meaning that only the harmonic parts of $\tilde{\mathbb{D}}$ undergo an eventual contraction, direction inversion and an equal reorientation with $\hat{\boldsymbol{Q}}$. The result Eq. (2.79) is the representation of the orientation average based on a general CCODF in terms of its central orientation and its texture eigenvalues λ_α up to $\alpha = 4 = r$. Higher-order texture eigenvalues have no influence on this orientation average. This representation of the orientation average based on CCODFs is actually already visible through the result Eq. (2.76). The advantage of the results Eq. (2.76) and Eq. (2.79) is that without further specification of the actual CCODF, the *relevant influence of all possible CCODFs* on the orientation average can be investigated by varying the central orientation and the *relevant texture eigenvalues* in their respective subspace of Λ, see Eq. (2.40). This representation offers a *finite and low dimensional* parametrization of the orientation average based on CCODFs *for all anisotropy classes* in terms of variables (texture eigenvalues) belonging to a *convex set*. These properties of the given representation might be attractive for material design purposes for polycrystalline materials and other optimization applications.

If a CODF $f(\boldsymbol{Q})$ is to be estimated with a convex combination of CCODFs as in Eq. (2.32), then the orientation average $\langle f, \mathbb{D}\rangle$ evaluates to

$$\langle f, \mathbb{D}\rangle = \sum_{\beta=1}^{n_c} f_\beta \langle g_\beta, \mathbb{D}\rangle \ , \tag{2.80}$$

where the orientation average of the β mode $\langle g_\beta, \mathbb{D}\rangle$ depends on the corresponding β central orientation $\hat{\boldsymbol{Q}}_\beta$ and β texture eigenvalues $(\lambda_\alpha)_\beta$.

For minor *and major* symmetric tensors $\tilde{\mathbb{D}}$, $\tilde{\boldsymbol{h}} = \boldsymbol{o}$, $\tilde{\boldsymbol{H}}'_3 = \boldsymbol{0}$ and $\tilde{\boldsymbol{H}}'_{\langle 3\rangle} = \boldsymbol{0}_{\langle 3\rangle}$ hold. This case implies that the relevant influence of a CCODF on $\langle g, \mathbb{D}\rangle$ is reflected by the reorientation with $\hat{\boldsymbol{Q}}$ and by the texture eigenvalues $(\lambda_2, \lambda_4) \in \Lambda_{2,4}$, see Eq. (2.39) and Fig. 2.1b. This low dimensional representation and the convex set $\Lambda_{2,4}$ are relevant for the homogenization of linear elastic properties of Chapter 3.

2.5.3 Orientation average based on the TFE

Representation in terms of texture coefficients. As briefly motivated in Section 2.5.1, the orientation average of an arbitrary r-th-order tensor $\tilde{\mathbb{D}}_{\langle r\rangle}$ can be compactly investigated based on the TFE. Consider the general harmonic decomposition Eq. (2.66) with the normalized reference tensors $\mathbb{H}'_{\langle\alpha\rangle\beta}$ as basis tensors for the respective harmonic spaces

$$\begin{aligned} \mathbb{D}_{\langle r\rangle}(\boldsymbol{Q}) = \boldsymbol{Q} \star \tilde{\mathbb{D}}_{\langle r\rangle} &= \sum_{\alpha=0}^{r} \sum_{\gamma=1}^{n_{r\alpha}} \mathbb{J}_{\langle r+\alpha\rangle\gamma} \left[\sum_{\beta=1}^{n'_\alpha} \tilde{h}_{\alpha\gamma\beta} \boldsymbol{Q} \star \mathbb{H}'_{\langle\alpha\rangle\beta} \right] , \\ \tilde{\mathbb{H}}'_{\langle\alpha\rangle\gamma} &= \sum_{\beta=1}^{n'_\alpha} \tilde{h}_{\alpha\gamma\beta} \mathbb{H}'_{\langle\alpha\rangle\beta} . \end{aligned} \tag{2.81}$$

The orientation average is transferred to the harmonic parts, or equally, to tensor functions $\mathbb{F}'_{\langle\alpha\rangle\beta}(\boldsymbol{Q}) = \boldsymbol{Q} \star \mathbb{H}'_{\langle\alpha\rangle\beta}$. The orientation average $\langle f, \mathbb{D}_{\langle r\rangle}\rangle$ based on an arbitrary r-th-order tensor $\tilde{\mathbb{D}}_{\langle r\rangle}$ is expressed in terms of the texture coefficients of the CODF $f(\boldsymbol{Q})$ by

$$\langle f, \mathbb{D}_{\langle r\rangle}\rangle = \sum_{\alpha=0}^{r} \sum_{\gamma=1}^{n_{r\alpha}} \mathbb{J}_{\langle r+\alpha\rangle\gamma} \left[\sum_{\beta=1}^{n'_\alpha} \tilde{h}_{\alpha\gamma\beta} \mathbb{V}'_{\langle\alpha\rangle\beta} \right] . \tag{2.82}$$

This representation of the orientation average equals the one given in Ganster and Gems (1985) and Guidi et al. (1992) but shows explicitly the different parts influencing the orientation average. The influence of the single crystal material is reflected by the scalars $\tilde{h}_{\alpha\gamma\beta}$ while the relevant influence of the crystallographic texture is reflected by the texture coefficients up to r-th-order. Higher-order texture coefficients of the CODF do not affect the orientation average. The representation is naturally linear in the single crystal behavior and in the texture.

The representation Eq. (2.82) offers, compared to the full function space of CODFs, a *finite and low dimensional* representation of the orientation average in terms of the texture coefficients belonging to the *convex* set $\mathcal{V}$, see Eq. (2.49). For the result Eq. (2.82) the subspace $\mathcal{V}_r$ suffices for the exploration of the influence of all CODFs on the orientation average $\langle f, \mathbb{D}_{\langle r\rangle}\rangle$. These properties make the result Eq. (2.82) attractive for applications and optimizations over the respective finite-dimensional convex sets.

For a minor symmetric fourth-order single crystal property $\tilde{\mathbb{D}}$, $\langle f, \mathbb{D}\rangle$ is represented explicitly as

$$
\begin{aligned}
\langle f, \mathbb{D}\rangle &= \mathrm{hd}\Big(\tilde{h}_{I1}, \tilde{h}_{I2}, \sum_{\beta=1}^{n_1'} \tilde{h}_{11\beta}\mathbb{V}'_{\langle 1\rangle\beta}, \sum_{\beta=1}^{n_2'} \tilde{h}_{21\beta}\mathbb{V}'_{\langle 2\rangle\beta}, \sum_{\beta=1}^{n_2'} \tilde{h}_{22\beta}\mathbb{V}'_{\langle 2\rangle\beta}, \\
&\qquad \sum_{\beta=1}^{n_2'} \tilde{h}_{23\beta}\mathbb{V}'_{\langle 2\rangle\beta}, \sum_{\beta=1}^{n_3'} \tilde{h}_{31\beta}\mathbb{V}'_{\langle 3\rangle\beta}, \sum_{\beta=1}^{n_4'} \tilde{h}_{41\beta}\mathbb{V}'_{\langle 4\rangle\beta}\Big)
\end{aligned} \tag{2.83}
$$

with

$$
\begin{aligned}
\tilde{\boldsymbol{h}} &= \sum_{\beta=1}^{n_1'} \tilde{h}_{11\beta}\mathbb{H}'_{\langle 1\rangle\beta}\,, \quad \tilde{\boldsymbol{H}}'_1 = \sum_{\beta=1}^{n_2'} \tilde{h}_{21\beta}\mathbb{H}'_{\langle 2\rangle\beta}\,, \quad \tilde{\boldsymbol{H}}'_2 = \sum_{\beta=1}^{n_2'} \tilde{h}_{22\beta}\mathbb{H}'_{\langle 2\rangle\beta}\,, \\
\tilde{\boldsymbol{H}}'_3 &= \sum_{\beta=1}^{n_2'} \tilde{h}_{23\beta}\mathbb{H}'_{\langle 2\rangle\beta}\,, \quad \tilde{\boldsymbol{H}}'_{\langle 3\rangle} = \sum_{\beta=1}^{n_3'} \tilde{h}_{31\beta}\mathbb{H}'_{\langle 3\rangle\beta}\,, \quad \tilde{\mathbb{H}}' = \sum_{\beta=1}^{n_4'} \tilde{h}_{41\beta}\mathbb{H}'_{\langle 4\rangle\beta}\,.
\end{aligned} \tag{2.84}
$$

For minor *and major* symmetric $\tilde{\mathbb{D}}$, $\tilde{\boldsymbol{h}} = \boldsymbol{o}$, $\tilde{\boldsymbol{H}}'_3 = \mathbf{0}$ and $\tilde{\boldsymbol{H}}'_{\langle 3\rangle} = \mathbf{0}_{\langle 3\rangle}$ hold. Therefore, for minor and major symmetric $\tilde{\mathbb{D}}$ its orientation average is described explicitly only depending on second- and fourth-order texture coefficients. This case will be important for many expressions in Chapter 3.

Connection to CCODFs via texture coefficients. The Dirichlet kernels $D_r(\omega)$ and associated functions $\mu_r(\omega)$ fulfill the identity

$$\int_{S_2} \mu_r(\omega)\mathrm{d}n = \int_{S_2} \mathbb{H}'_{\langle r\rangle} \cdot (\boldsymbol{Q} \star \mathbb{H}'_{\langle r\rangle})\mathrm{d}n \tag{2.85}$$

for arbitrary normalized harmonic reference tensor $\mathbb{H}'_{\langle r\rangle}$. This property implies that the orientation average of an arbitrary harmonic r-th-order tensor $\mathbb{H}'_{\langle r\rangle}$ with $\mathbb{F}'_{\langle r\rangle}(\boldsymbol{Q}) = \boldsymbol{Q} \star \mathbb{H}'_{\langle r\rangle}$ based on a CCODF evaluates to

$$\langle g, \mathbb{F}'_{\langle r\rangle}\rangle = \hat{\boldsymbol{Q}} \star (\lambda_r \mathbb{H}'_{\langle r\rangle}) \,. \tag{2.86}$$

This result is shown as follows. The general orientation average of a harmonic tensor $\mathbb{H}'_{\langle r\rangle}$ delivers due to linearity also a harmonic tensor $\langle f, \mathbb{F}'_{\langle r\rangle}\rangle$. This also holds for the orientation average based on a CCODF centered around the identity, which might be represented as $\langle g, \mathbb{F}'_{\langle r\rangle}\rangle|_{\hat{\boldsymbol{Q}}=\boldsymbol{I}} = \lambda_r \mathbb{H}'_{\langle r\rangle} + \mathbb{R}'_{\langle r\rangle}$, with a harmonic tensor $\mathbb{R}'_{\langle r\rangle}$ representing the eventual non-vanishing rest. The rest $\mathbb{R}'_{\langle r\rangle}$, if it does not vanish, has to be a linear function of $\mathbb{H}'_{\langle r\rangle}$, say, $\mathbb{R}'_{\langle r\rangle} = \mathbb{L}'_{\langle 2r\rangle}[\mathbb{H}'_{\langle r\rangle}]$, where $\mathbb{L}'_{\langle 2r\rangle}$ has to map from harmonic into harmonic tensors. The tensor $\mathbb{R}'_{\langle r\rangle}$ would have to be orthogonal to $\mathbb{H}'_{\langle r\rangle}$ since $\lambda_r(\mathbb{H}'_{\langle r\rangle} \cdot \mathbb{H}'_{\langle r\rangle}) = \mathbb{H}'_{\langle r\rangle} \cdot \langle g, \mathbb{F}'_{\langle r\rangle}\rangle|_{\hat{\boldsymbol{Q}}=\boldsymbol{I}}$ holds for arbitrary $\mathbb{H}'_{\langle r\rangle}$ due to Eq. (2.85). This implies $\mathbb{H}'_{\langle r\rangle} \cdot \mathbb{L}'_{\langle 2r\rangle}[\mathbb{H}'_{\langle r\rangle}] = 0$ for arbitrary $\mathbb{H}'_{\langle r\rangle}$, which can only be fulfilled if $\mathbb{L}'_{\langle 2r\rangle}$ maps every harmonic tensor into the zero tensor. This implies that the tensor $\mathbb{R}'_{\langle r\rangle}$ vanishes.

If the general CODF is represented as a convex combination of CCODFs, as in Eq. (2.32), then the general harmonic decomposition and the result Eq. (2.86) allow to express the general orientation average $\langle f, \mathbb{D} \rangle$ depending on the central orientations and texture eigenvalues of the CCODFs. Due to the properties of the harmonic decomposition, the general orientation average is transferred to the harmonic parts which can be represented as linear combinations of the normalized reference tensors, as in Eq. (2.84). The general orientation average of the normalized reference tensors equals per definition the texture coefficients and can be expressed under these considerations using Eq. (2.86) as

$$\mathbb{V}'_{\langle\alpha\rangle\beta} = \langle f, \mathbb{F}'_{\langle\alpha\rangle\beta} \rangle = \sum_{\gamma=1}^{n_c} f_\gamma \langle g_\gamma, \mathbb{F}'_{\langle\alpha\rangle\beta} \rangle = \sum_{\gamma=1}^{n_c} f_\gamma \hat{\boldsymbol{Q}}_\gamma \star ((\lambda_\alpha)_\gamma \mathbb{H}'_{\langle\alpha\rangle\beta}) \, . \tag{2.87}$$

This can be applied to the general orientation average of arbitrary order tensors with arbitrary symmetries and helps to connect more clearly the results Eq. (2.79), Eq. (2.80) and Eq. (2.83) for fourth-order minor symmetric tensors. Alternatively, the result Eq. (2.87) offers a more trackable parametrization of texture coefficients as a truncated series regarding central orientations and texture eigenvalues belonging to the set Λ. This may be used together with the motivated set $\check{\mathcal{V}}^{\mathrm{sym}}$, see Eq. (2.54), to delineate subsets of the set of all possible texture coefficients $\mathcal{V}$ for symmetry groups of interest. As an example, consider polycrystals of hexagonal materials with aligned macroscopic hexagonal symmetry, as discussed in Section 2.3.3 and depicted in Fig. 2.2. If only one CCODF is used in Eq. (2.87) with $\hat{\boldsymbol{Q}} = \boldsymbol{I}$, for the micro-macro-hexagonal case $V'_{11} = \lambda_2/\sqrt{6}$ and $V'_{1111} = 3\lambda_4/(2\sqrt{70})$ hold. The symmetry group independent set $\Lambda_{2,4}$ is rescaled accordingly, as depicted in green in Fig. 2.3. As expected, the set $\Lambda_{2,4}$ is a subset of $\mathcal{V}_4^{\mathrm{hex}}$ and covers already with a single CCODF a large part of $\mathcal{V}_4^{\mathrm{hex}}$ ($\approx 59\%$).

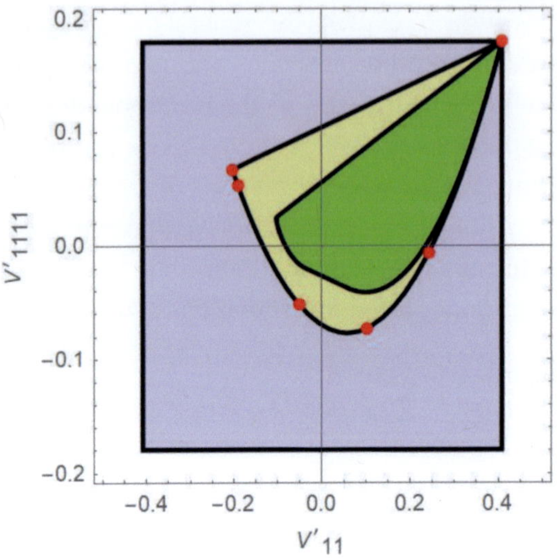

Figure 2.3: Rescaled set $\Lambda_{2,4}$ in green together with the sets $\mathcal{V}_4^{\text{hex}}$ (light green region), $\check{\mathcal{V}}_4^{\text{hex}}$ (convex combination of the red points) and $\mathcal{N}_4^{\text{hex}}$ (blue region)

2.5.4 Linearization with respect to texture

It should be remarked that, if only weak texture changes concerning a reference CODF $f_r(\boldsymbol{Q})$ are visible in experiments or certain applications, then a linearization of several physical quantities with respect to the texture might be considered. For example, the second-order tensor

$$\boldsymbol{A} = \int_v h(\boldsymbol{x})\boldsymbol{\alpha}(\boldsymbol{x})\mathrm{d}v \tag{2.88}$$

is considered as a functional of the scalar field $h(\boldsymbol{x})$ with given second-order tensor field $\boldsymbol{\alpha}(\boldsymbol{x})$ over a material domain v over the three-dimensional physical space. For a compact notation, $\boldsymbol{A} = \boldsymbol{A}(h)$ and $\boldsymbol{A}_r = \boldsymbol{A}(h_r)$ for a reference scalar function $h_r(\boldsymbol{x})$ will be used. The variation $\delta\boldsymbol{F}$ and variational derivative $(\delta\boldsymbol{F}/\delta h)(g)$ are defined as

$$\delta \boldsymbol{F} = \left. \frac{\mathrm{d}\boldsymbol{F}(h + \epsilon \delta h)}{\mathrm{d}\epsilon} \right|_{\epsilon = 0} = \frac{\delta \boldsymbol{F}}{\delta h}(\delta h) \,, \tag{2.89}$$

where the variational derivative is linear in its argument. Trivially, for the functional $\boldsymbol{A}$ hold

$$\delta \boldsymbol{A} = \int_v \delta h(\boldsymbol{x}) \boldsymbol{\alpha}(\boldsymbol{x}) \mathrm{d}v \,, \quad \frac{\delta \boldsymbol{A}}{\delta h}(g_1 + g_2) = \int_v (g_1(\boldsymbol{x}) + g_2(\boldsymbol{x})) \boldsymbol{\alpha}(\boldsymbol{x}) \mathrm{d}v \,. \tag{2.90}$$

The variation of the inverse is obtained as $\delta \boldsymbol{A}^{-1} = (-1)\boldsymbol{A}^{-1}(\delta \boldsymbol{A})\boldsymbol{A}^{-1}$, such that the variational linearization of the inverse of $\boldsymbol{A}$ with respect to $h(\boldsymbol{x})$ near a scalar function $h_r(\boldsymbol{x})$ is expressed as

$$\begin{aligned} \boldsymbol{A}^{-1} &\approx \boldsymbol{A}_r^{-1} + \left. \tfrac{\delta \boldsymbol{A}^{-1}}{\delta h} \right|_{h = h_r} (h - h_r) \\ &= \boldsymbol{A}_r^{-1} - \boldsymbol{A}_r^{-1} \left(\tfrac{\delta \boldsymbol{A}}{\delta h}(h - h_r) \right) \boldsymbol{A}_r^{-1} \\ &= \boldsymbol{A}_r^{-1} - \boldsymbol{A}_r^{-1} (\boldsymbol{A} - \boldsymbol{A}_r) \boldsymbol{A}_r^{-1} \\ &= 2\boldsymbol{A}_r^{-1} - \boldsymbol{A}_r^{-1} \boldsymbol{A} \boldsymbol{A}_r^{-1} \,. \end{aligned} \tag{2.91}$$

In this work, bounds and approximations for the effective fourth-order stiffness tensor $\bar{\mathbb{C}}$ of multiphase polycrystalline materials will be considered. The expression given in Eq. (2.91), see also Lobos Fernández and Böhlke (2018), might be used for linearization of several expressions concerning nonlinear expressions in terms of the stiffness depending on the CODF (e.g., the Reuss average and HS bounds for elastic properties of polycrystals, see Reuss (1929), Hashin and Shtrikman (1962), Willis (1977)) or other density functions. For example, the inverse of the general orientation average might be linearized with respect to a given reference CODF $f_r(\boldsymbol{Q})$ as follows

$$(\langle f, \mathbb{D}_{\langle r \rangle} \rangle)^{-1} \approx 2(\langle f_r, \mathbb{D}_{\langle r \rangle} \rangle)^{-1} - (\langle f_r, \mathbb{D}_{\langle r \rangle} \rangle)^{-1} (\langle f, \mathbb{D}_{\langle r \rangle} \rangle)(\langle f_r, \mathbb{D}_{\langle r \rangle} \rangle)^{-1} \,. \tag{2.92}$$

All orientation averages can be treated as presented in this chapter and represented in terms of texture coefficients or eigenvalues. The expression Eq. (2.92) is naturally linear in the relevant texture coefficients of $f(\boldsymbol{Q})$. But it should be noted, that, as every linearization, limited applicability is to be expected. This is due to two points of view. First, depending on the reference texture (and corresponding texture coefficients), it is not clear how far from the reference texture a sensible answer is to be expected. It has to be considered where the reference texture is found in the set $\mathcal{V}$ of all texture coefficients. And second, how far from the reference texture a sensible answer is to be expected will be influenced by the strength of the harmonic parts of the physical quantity $\tilde{\mathbb{D}}_{\langle r \rangle}$. Depending on, e.g., the norm of each harmonic part, the influence of different texture coefficients will be notable or negligible, independently of the reference texture.

Nevertheless, the linearization with respect to the crystallographic texture may be a valuable and efficient tool for applications with given and well investigated material properties and a reference texture at a "sufficiently safe" position within $\mathcal{V}$. Such a position could be the isotropic case, i.e., $f(\boldsymbol{Q}) = 1$. The expression Eq. (2.92) simplifies significantly. All reference texture coefficients vanish, only the isotropic part of $\tilde{\mathbb{D}}_{\langle r \rangle}$ remains and the weak texture expressions of Man and Huang (2011), Man and Huang (2012) and Du and Man (2017) are immediately obtainable.

Alternatively, the linearization with respect to the texture may be useful for optimization problems. For nonlinear expressions in the texture trying to minimize an objective function depending on the texture, classical Newton algorithms will require the linearization. For this kind of applications, the just discussed linearizations are expected to be of some use.

2.6 Review

In this chapter, the basic calculus of orientations $\boldsymbol{Q} \in SO(3) \subset Orth$ has been motivated. The CODF $f(\boldsymbol{Q})$ and the CCODF $g(\hat{\omega})$ have been presented, together with the texture eigenvalues λ_α and texture coefficients $\mathbb{V}'_{\langle\alpha\rangle\beta}$ as coefficients of the respective Fourier expansions. The sets of all possible texture eigenvalues Λ and texture coefficients $\mathcal{V}$ have been defined. The convexity and importance of these sets have been discussed, together with the instructive subset $\check{\mathcal{V}}^{\text{sym}}_\alpha$ based on a finite number of single crystal states and group symmetrization up to tensor order α. The orientation average of general r-th-order tensor quantities $\tilde{\mathbb{D}}_{\langle r\rangle}$ has been deduced based on the CCODF in terms of texture eigenvalues and based on the CODF in terms of texture coefficients through the harmonic decomposition. The case for a fourth-order minor and major symmetric tensor has been given explicitly as a preparation for the expressions in the upcoming chapters. Linearizations with respect to the texture have been discussed.

Chapter 3

Homogenization of linear and nonlinear properties of multiphase polycrystalline materials

3.1 Overview

This chapter begins with the basic probabilistic description of random heterogeneous materials and the basic idea of homogenization of these materials. Then, the homogenization of the classical linear elastic mechanical problem with eigenfields is reviewed. In the present work, zeroth-, first- and second-order bounds of multiphase polycrystalline materials are discussed. The explicit representation of the second-order Hashin-Shtrikman bounds in terms of material phase concentrations and texture coefficients is derived. These expressions are then used for the generation of bounds and pragmatic approximations for optimization problems in the field of materials design. Approximations for the thermomechanical case based on the same set of variables (material volume fractions and texture coefficients) are presented. The applicability of the obtained expressions for physically nonlinear bounds based on linear comparison composites is discussed briefly.

3.2 Probabilistic description and homogenization of random inhomogeneous materials

Random inhomogeneous material. In the present work, following Willis (1981), Torquato (2002) and Milton (2002), material bodies with material inhomogeneities are considered for stochastic manufacturing processes. The inhomogeneities are characterized by a length l_μ, being greater than the maximum size of all inhomogeneities. The material properties vary in physical space at this microscopic length scale l_μ which is assumed significantly smaller than the length scale l_M of the body to be described by the material region Ω. The length scale l_M is referred to as the macroscopic length scale. From a macroscopic point of view, at some intermediate length scale l_m, referred to as the mesoscale, the inhomogeneous body might appear homogeneous and macroscopic fields (e.g., mechanical stress and strain) might vary slowly in space with respect to l_m. These three fundamental scales are assumed to fulfill $l_\mu \ll l_m \ll l_M$, which is referred to as a scale separation. For example, for a typical piece of metal, the macroscopic and microscopic length scales take values around $l_M \approx 1m - 10^{-2}m$ and $l_\mu \approx 10^{-4}m - 10^{-6}m$. Scale separation is assumed throughout the present work. The considered material bodies might be produced by complex stochastic manufacturing processes, such that from sample to sample the material arrangement might change with certain probabilities. These materials are referred to as random heterogeneous materials.

Ensemble. An ensemble is the collection of all samples α from a sample space $\mathcal{A}$. From a manufacturing point of view, every realization of a heterogeneous material is a sample α from all possible realizations of the considered random manufacturing process. The different realizations and corresponding microstructures are reproduced with a certain probability depending on the manufacturing parameters. The ensemble average of a physical quantity $\phi(\alpha)$ (e.g., the stiffness, the thermal expansion, or other material property) varying from sample to sample is defined as

$$\langle \phi \rangle = \int_{\mathcal{A}} \phi(\alpha) p(\alpha) \mathrm{d}\alpha \,, \tag{3.1}$$

where p denotes the probability measure of the sample space.

Description of microstructures. It is assumed that all samples occupy the same finite region Ω in the three-dimensional real space. Each sample of the ensemble might show complex material distribution. For each sample, each material constituent is referred to as a (material) phase. It is remarked, that for a single constituent polycrystalline material, every single crystalline orientation is to be considered as a distinct phase, as discussed in Section 1.1.

The microstructure, i.e., the material distribution of the m phases of the sample, is characterized by the indicator function $I^i(\boldsymbol{x}, \alpha)$ of phase i, $i = 1, 2, \ldots, m$, defined as

$$I^i(\boldsymbol{x}, \alpha) = \begin{cases} 1 & \boldsymbol{x} \in \Omega_i \subset \Omega \\ 0 & \text{else} \end{cases} \,. \tag{3.2}$$

The material region Ω_i refers to the region of phase i in sample α. The indicator functions have projector properties, i.e., $I^i(\boldsymbol{x}, \alpha) I^j(\boldsymbol{x}, \alpha) = 0$ $\forall i \neq j$, $I^i(\boldsymbol{x}, \alpha) I^i(\boldsymbol{x}, \alpha) = I^i(\boldsymbol{x}, \alpha)$ and $\sum_{i=1}^{m} I^i(\boldsymbol{x}, \alpha) = 1$.

A physical quantity, e.g., the position dependent stiffness $\mathbb{C}(\boldsymbol{x}, \alpha)$ in sample α is described in this work for a random medium composed of m phases with phasewise constant material properties as

$$\mathbb{C}(\boldsymbol{x}, \alpha) = \sum_{i=1}^{m} \mathbb{C}_i I^i(\boldsymbol{x}, \alpha) \,. \tag{3.3}$$

For the present work, the ensemble average of several physical quantities is of interest. For example, the ensemble average of the stiffness at two different points is expressed as

$$\begin{aligned}\langle \mathbb{C}\mathbb{C}\rangle(\boldsymbol{x}_1, \boldsymbol{x}_2) &= \int_{\mathcal{A}} \mathbb{C}(\boldsymbol{x}_1, \alpha)\mathbb{C}(\boldsymbol{x}_2, \alpha)p(\alpha)\mathrm{d}\alpha \\ &= \sum_{i_1=1}^{m}\sum_{i_2=1}^{m} \mathbb{C}_{i_1}\mathbb{C}_{i_2} \int_{\mathcal{A}} I^{i_1}(\boldsymbol{x}_1, \alpha)I^{i_2}(\boldsymbol{x}_2, \alpha)p(\alpha)\mathrm{d}\alpha\,. \end{aligned} \tag{3.4}$$

It is obvious that the value of this expression is controlled by the geometrical arrangement and the probability of the different phases in the random material. This motivates the necessity to define probability functions for different phases.

The n-point probability functions (n-PPFs). For the probabilistic description of microstructures, correlations between single and multiple phases are of interest. The functions

$$\begin{aligned} S_n^{i_1 \dots i_n}(\boldsymbol{x}_1, \dots, \boldsymbol{x}_n) &= \langle I^{i_1} \dots I^{i_n}\rangle(\boldsymbol{x}_1, \dots, \boldsymbol{x}_n) \\ &= \int_{\mathcal{A}} \prod_{j=1}^{n} I^{i_j}(\boldsymbol{x}_j, \alpha)p(\alpha)\mathrm{d}\alpha \end{aligned} \tag{3.5}$$

are referred to as the n-point probability functions (n-PPFs), see, e.g., Willis (1981) or Torquato (2002). These functions reflect the relative amount of times (the probability) a specific event happens in the ensemble. For example, the two-point auto-correlation of a phase i, $S_2^{ii}(\boldsymbol{x}_1, \boldsymbol{x}_2) = S_2^{i}(\boldsymbol{x}_1, \boldsymbol{x}_2)$ for simplicity, reflects the probability that phase i is found at the same time at positions $\boldsymbol{x}_1$ and $\boldsymbol{x}_2$ with respect to the ensemble. For $\boldsymbol{x}_1 = \boldsymbol{x}_2$, the auto-correlation simplifies to the one-point probability function $S^i(\boldsymbol{x}_1)$ which reflects the volume fraction of phase i at $\boldsymbol{x}_1$. This naturally means, that the n-PPFs for given n contain all smaller $\tilde{n}$-PPFs for $\tilde{n} < n$ as special cases.

Statistical homogeneity and isotropy. Due to manufacturing conditions, some processes might show, concerning the ensemble average of a physical quantity, no change from position to position within Ω. This is referred to as statistical homogeneity and is defined as

$$S_n^{i_1 \dots i_n}(\boldsymbol{x}_1, \dots, \boldsymbol{x}_n) = S_n^{i_1 \dots i_n}(\boldsymbol{x}_1 + \boldsymbol{y}, \dots, \boldsymbol{x}_n + \boldsymbol{y}) \quad \forall \boldsymbol{y} \in \Omega_y \; \forall n \, , \tag{3.6}$$

where Ω_y denotes the set of all possible displacements such that $\boldsymbol{x}_i + \boldsymbol{y} \in \Omega$. From a theoretical point of view, for this to make sense, Ω and Ω_y have to be extended to infinity.

If an ensemble is statistically homogeneous, then the n-PPFs of the ensemble can be reduced by one argument, e.g., by choosing $\boldsymbol{y} = -\boldsymbol{x}_1$, i.e., $S^{i_1 \dots i_n}(\Delta \boldsymbol{x}_2, \dots, \Delta \boldsymbol{x}_n)$ with $\Delta \boldsymbol{x}_i = \boldsymbol{x}_i - \boldsymbol{x}_1, i \geq 2$. It should be emphasized, that naturally other choices for $\boldsymbol{y}$ in case of statistically homogeneity are possible since for these cases the n-PPFs have no preferred origin (translation invariance).

Further, a statistically homogeneous ensemble is referred to as statistically isotropic, if the n-PPFs are rotationally invariant, i.e.,

$$S_n^{i_1 \dots i_n}(\Delta \boldsymbol{x}_2, \dots, \Delta \boldsymbol{x}_n) = S_n^{i_1 \dots i_n}(\boldsymbol{Q} \Delta \boldsymbol{x}_2, \dots, \boldsymbol{Q} \Delta \boldsymbol{x}_n) \; \forall \boldsymbol{Q} \in SO(3) \; \forall n \tag{3.7}$$

holds, see Torquato (2002). For such material ensembles, it follows that the n-PPFs depend not on the direction of the arguments but only on their magnitudes $\|\Delta \boldsymbol{x}_i\|$.

Ergodic hypothesis. All possible states of a random material system are contained in an ensemble. The ensemble average reflects the average state of the sample. If the sample is statistically homogeneous, then this average state is found at every point in a probabilistic sense. One may imagine now an infinite realization of the random material containing all states of the ensemble. Averaging over such an infinite volume would also average all possible states of the sample and deliver the same value as the ensemble average of the statistically homogeneous medium. This is the ergodic hypothesis: the complete probabilistic information of a statistically homogeneous medium is equally obtained through the volume average over an infinite representative volume element (RVE) with material region v

$$\begin{aligned} & S_n^{i_1 i_2 \dots i_n}(\Delta \boldsymbol{x}_2, \dots, \Delta \boldsymbol{x}_n) \\ &= \langle I^{i_1} I^{i_2} \dots I^{i_n} \rangle (\Delta \boldsymbol{x}_2, \dots, \Delta \boldsymbol{x}_n) \\ &= \lim_{v \to \infty} \frac{1}{v} \int_v I^{i_1}(\boldsymbol{x}) I^{i_2}(\boldsymbol{x} + \Delta \boldsymbol{x}_2) \dots I^{i_n}(\boldsymbol{x} + \Delta \boldsymbol{x}_n) \mathrm{d}v \\ &= \langle I^{i_1} I^{i_2} \dots I^{i_n} \rangle_v (\Delta \boldsymbol{x}_2, \dots, \Delta \boldsymbol{x}_n) \,, \end{aligned} \tag{3.8}$$

see, e.g., Beran (1968) or Torquato (2002). Based on the ergodic hypothesis, the ensemble average of a physical quantity Eq. (3.1) for statistically homogeneous media is computed as the volume average over one specific realization, i.e., an RVE, and is denoted in this work as $\langle \phi \rangle = \langle \phi \rangle_v$.

Homogenization The notion of a finite RVE is related to the notion of a mesoscale for statistical homogeneous media. The dimensions of the finite RVE is to be chosen as $\sqrt[3]{v} \approx l_m$. This means that, roughly speaking, the finite RVE has to be large enough compared to the microstructure such that volume averaging over v contains enough *representative* statistical information. Scale separation and the consideration of an RVE allow examining elliptic differential equations of the type

$$\mathrm{div}(\boldsymbol{A}\,\mathrm{grad}(\phi)) = \boldsymbol{\rho} \,, \quad \boldsymbol{x} \in \Omega \,, \tag{3.9}$$

with bounded positive definite tensor field $\boldsymbol{A}$, representing material phase properties, a potential function ϕ, and a inhomogeneity $\boldsymbol{\rho}$, see, e.g., Milton (2002). The material properties $\boldsymbol{A}$ oscillate rapidly over the macroscopic body Ω, and the solution of this kind of equation might be highly cumbersome if the complete microstructure is to be taken into account. From a physical point of view, if the material properties vary so rapidly over the macroscopic body and macroscopic physical state $\langle \text{grad}(\phi) \rangle_v$ varies slowly over the mesoscale, then it is sensible to seek for effective material properties $\bar{\boldsymbol{A}}$ satisfying

$$\begin{aligned} \text{div}(\bar{\boldsymbol{A}} \, \langle \text{grad}(\phi) \rangle_v) &= \boldsymbol{\rho} \,, \quad \boldsymbol{x} \in \Omega \,, \\ \langle \boldsymbol{A} \, \text{grad}(\phi) \rangle_v &= \bar{\boldsymbol{A}} \, \langle \text{grad}(\phi) \rangle_v \end{aligned} \tag{3.10}$$

This is the basic idea of homogenization. The determination of the homogenized or equivalently effective material properties $\bar{\boldsymbol{A}}$ of the RVE is one of the main objectives treated in the wide literature of homogenization theory. Also, bounds and sensible approximations of the effective material properties of the RVE are of great practical use for materials science. From the perspective of materials design, it is highly desired to obtain finite and low dimensional representations of the effective properties, corresponding bounds and/or approximations. This kind of representation facilitates substantially the optimization problems posed in materials design in which material properties are tailored corresponding to prescribed application constraints, see, e.g., Adams et al. (2013) and Kalidindi (2015).

3.3 Homogenization with eigenfields

Linear elastic boundary value problem with eigenfields. For linear elastic problems taking into account eigenfields, the equilibrium condition at a material point $\boldsymbol{x} \in \Omega$ of each material sample $\alpha \in \mathcal{A}$ with homogeneous boundary conditions and corresponding elastic law for small deformations are formulated as

$$\operatorname{div}(\boldsymbol{\sigma}) = 0 \quad \boldsymbol{x} \in \Omega\,, \quad \boldsymbol{u} = \boldsymbol{\varepsilon}_0 \boldsymbol{x} \quad \boldsymbol{x} \in \partial\Omega\,, \quad \boldsymbol{\sigma} = \mathbb{C}[\boldsymbol{\varepsilon}] - \boldsymbol{s}\,. \tag{3.11}$$

The Cauchy stress tensor is denoted as $\boldsymbol{\sigma}$ and is determined by the material law $\boldsymbol{\sigma} = \mathbb{C}[\boldsymbol{\varepsilon}] - \boldsymbol{s}$. The infinitesimal strain tensor is denoted as $\boldsymbol{\varepsilon} = \operatorname{sym}(\operatorname{grad}(\boldsymbol{u}))$, the arbitrary stress eigenfield $\boldsymbol{s}$ and the material stiffness $\mathbb{C}$. It is assumed, that $\boldsymbol{s}$ is independent of $\boldsymbol{\varepsilon}$. All tensor fields are position and sample dependent, in general. Only the boundary strain $\boldsymbol{\varepsilon}_0$ is assumed constant. It should be remarked that the boundary conditions of (3.11) do not allow for displacement fluctuations on the boundary of Ω. In this work, it is assumed that the material region Ω has no pores or cracks, such that $\boldsymbol{\varepsilon}_0 = \int_\Omega \boldsymbol{\varepsilon} \mathrm{d}v/\Omega$ holds. Each material sample α is assumed to occupy the same volume Ω, the material distribution, i.e., $\mathbb{C}(\boldsymbol{x}, \alpha)$, might change from sample to sample.

Solution for single sample. The solution of linear elastic problems with eigenfields can be obtained formally following Willis (1981) by using a constant comparison material with stiffness $\mathbb{C}_0$ and the exact polarization field $\hat{\boldsymbol{\tau}}$

$$\boldsymbol{\sigma} = \mathbb{C}_0[\boldsymbol{\varepsilon}] + \hat{\boldsymbol{\tau}}\,, \quad \hat{\boldsymbol{\tau}} = \hat{\mathbb{C}}[\boldsymbol{\varepsilon}] - \boldsymbol{s}\,, \quad \hat{\mathbb{C}} = \mathbb{C} - \mathbb{C}_0\,. \tag{3.12}$$

Based on Green's function, the solution of the boundary value problem for each sample is expressed with a sample dependent linear non-local operator $\mathbb{G}$, see Willis (1977; 1981), as

$$\boldsymbol{\varepsilon} = \boldsymbol{\varepsilon}_0 - \mathbb{G}\{\hat{\boldsymbol{\tau}}\}\,, \quad \mathbb{G}\{\hat{\boldsymbol{\tau}}\}(\boldsymbol{x}, \alpha) = \int_\Omega \mathbb{G}(\boldsymbol{x}, \boldsymbol{x}', \alpha)[\hat{\boldsymbol{\tau}}(\boldsymbol{x}', \alpha)]\mathrm{d}v'\,. \tag{3.13}$$

The solution might be reformulated in terms of $\boldsymbol{\varepsilon}$ by eliminating $\hat{\boldsymbol{\tau}}$ as

$$(\mathbb{I}^S + \mathbb{G}\hat{\mathbb{C}})\{\boldsymbol{\varepsilon}\} = \boldsymbol{\varepsilon}_0 + \mathbb{G}\{\boldsymbol{s}\}\,, \tag{3.14}$$

where $\mathbb{I}^S$ is temporarily used as the identity operator and the term $(\mathbb{I}^S + \mathbb{G}\hat{\mathbb{C}})\{\varepsilon\} = \varepsilon + \mathbb{G}\{\hat{\mathbb{C}}[\varepsilon]\}$ is to be read as

$$\varepsilon(\boldsymbol{x}, \alpha) + \int_\Omega \mathbb{G}(\boldsymbol{x}, \boldsymbol{x}', \alpha)[\hat{\mathbb{C}}(\boldsymbol{x}', \alpha)[\varepsilon(\boldsymbol{x}', \alpha)]]\mathrm{d}v' \,. \tag{3.15}$$

Due to the length of these and upcoming formal expressions, in this section we will only indicate the formal application of non-local operators. For each sample, the solution of the respective boundary value problem can be expressed formally due to the uniqueness of the solution as

$$\varepsilon = \mathbb{A}_0\{\varepsilon_0\} + \boldsymbol{a}_0 \,, \quad \mathbb{A}_0 = (\mathbb{I}^S + \mathbb{G}\hat{\mathbb{C}})^{-1} \,, \quad \boldsymbol{a}_0 = \mathbb{A}_0\{\mathbb{G}\{\boldsymbol{s}\}\} \,, \tag{3.16}$$

The non-local operator $\mathbb{A}_0$ and the field $\boldsymbol{a}_0$ fulfill the conditions

$$\frac{1}{\Omega}\int_\Omega \mathbb{A}_0 \mathrm{d}v = \mathbb{I}^S \,, \quad \frac{1}{\Omega}\int_\Omega \boldsymbol{a}_0 \mathrm{d}v = \boldsymbol{0} \,. \tag{3.17}$$

Solution in terms of ensemble average quantities. The ensemble average of a position dependent material quantity $q(\boldsymbol{x}, \alpha)$ is expressed as follows

$$\langle q \rangle(\boldsymbol{x}) = \int_\mathcal{A} q(\boldsymbol{x}, \alpha) p(\mathrm{d}\alpha) \,. \tag{3.18}$$

The deterministic displacement boundary conditions of (3.11) controlled by the constant ε_0 can be eliminated by consideration of the ensemble averaged fields

$$\varepsilon_0 = \langle \mathbb{A}_0 \rangle^{-1}\{\langle \varepsilon \rangle - \langle \boldsymbol{a}_0 \rangle\} \,. \tag{3.19}$$

The solution of the linear elastic problem for each sample is now reformulated as

$$\varepsilon = \mathbb{A}\{\langle \varepsilon \rangle\} + \boldsymbol{a} \,, \quad \mathbb{A} = \mathbb{A}_0 \langle \mathbb{A}_0 \rangle^{-1} \,, \quad \boldsymbol{a} = \boldsymbol{a}_0 - \mathbb{A}\{\langle \boldsymbol{a}_0 \rangle\} \,, \tag{3.20}$$

with the normalized operator $\mathbb{A}$, i.e., $\langle \mathbb{A} \rangle = \mathbb{I}^S$, and the fluctuation field $\boldsymbol{a}$, i.e., $\langle \boldsymbol{a} \rangle = \boldsymbol{0}$.

Effective material properties of ensemble. The effective material behavior is defined in terms of the ensemble average of the strain and stresses based on (3.20) as

$$\langle \boldsymbol{\sigma} \rangle = \bar{\mathbb{C}}\{\langle \boldsymbol{\varepsilon} \rangle\} - \bar{\boldsymbol{s}} \,, \quad \bar{\mathbb{C}} = \langle \mathbb{C}\mathbb{A} \rangle \,, \quad \bar{\boldsymbol{s}} = \langle \boldsymbol{s} - \mathbb{C}[\boldsymbol{a}] \rangle \,. \tag{3.21}$$

It should be noted that the fields $\boldsymbol{a}_0$, $\boldsymbol{a}$ and $\bar{\boldsymbol{s}}$ are all linear in the stress eigenfield $\boldsymbol{s}$.

Effective potential of ensemble. An energy density W is considered as a potential for the stresses, i.e., $\boldsymbol{\sigma} = \partial W / \partial \boldsymbol{\varepsilon}$, in the following forms

$$W = \frac{1}{2} \boldsymbol{\varepsilon} \cdot \mathbb{C}[\boldsymbol{\varepsilon}] - \boldsymbol{\varepsilon} \cdot \boldsymbol{s} - \frac{1}{2} k = \frac{1}{2} (\boldsymbol{\varepsilon} \cdot \boldsymbol{\sigma} - \boldsymbol{\varepsilon} \cdot \boldsymbol{s} - k) \,. \tag{3.22}$$

The ensemble average of W can be simplified assuming the validity of $\langle \boldsymbol{\varepsilon} \cdot \boldsymbol{\sigma} \rangle = \langle \boldsymbol{\varepsilon} \rangle \cdot \langle \boldsymbol{\sigma} \rangle$ and inserting the effective law for $\langle \boldsymbol{\sigma} \rangle$ as

$$\begin{aligned} 2\langle W \rangle &= \langle \boldsymbol{\varepsilon} \cdot \boldsymbol{\sigma} \rangle - \langle \boldsymbol{\varepsilon} \cdot \boldsymbol{s} \rangle - \langle k \rangle \\ &= \langle \boldsymbol{\varepsilon} \rangle \cdot \langle \boldsymbol{\sigma} \rangle - \langle (\mathbb{A}\{\langle \boldsymbol{\varepsilon} \rangle\} + \boldsymbol{a}) \cdot \boldsymbol{s} \rangle - \langle k \rangle \\ &= \langle \boldsymbol{\varepsilon} \rangle \cdot \bar{\mathbb{C}}\{\langle \boldsymbol{\varepsilon} \rangle\} - \langle \boldsymbol{\varepsilon} \rangle \cdot (\bar{\boldsymbol{s}} + \langle \mathbb{A}^{\dagger}\{\boldsymbol{s}\} \rangle) - \langle k + \boldsymbol{a} \cdot \boldsymbol{s} \rangle \,, \end{aligned} \tag{3.23}$$

where the dagger $\dagger$ denotes the adjoint operator. In this case, if the condition is imposed that the effective material behavior is also to be derivable from the effective energy

$$\frac{\partial \langle W \rangle}{\partial \langle \boldsymbol{\varepsilon} \rangle} = \langle \boldsymbol{\sigma} \rangle \quad \forall \langle \boldsymbol{\varepsilon} \rangle, \boldsymbol{s}, \tag{3.24}$$

then

$$\bar{\mathbb{C}} = \bar{\mathbb{C}}^{\dagger} \,, \qquad \bar{\boldsymbol{s}} = \langle \mathbb{A}^{\dagger}\{\boldsymbol{s}\} \rangle \tag{3.25}$$

hold, i.e., the effective stiffness is self-adjoint, and the effective eigenfield can be obtained by the alternative expression. Under these conditions, the effective elastic energy density is expressed in terms of the effective material properties as

$$\langle W \rangle = \frac{1}{2}\langle \varepsilon \rangle \cdot \bar{\mathbb{C}}\{\langle \varepsilon \rangle\} - \langle \varepsilon \rangle \cdot \bar{s} - \frac{1}{2}\bar{k} \tag{3.26}$$

with the effective material properties of the ensemble

$$\bar{\mathbb{C}} = \langle \mathbb{C}\mathbb{A} \rangle\,, \quad \bar{s} = \langle \mathbb{A}^{\dagger}\{s\} \rangle\,, \quad \bar{k} = \langle k \rangle + \langle a \cdot s \rangle\,. \tag{3.27}$$

These formal expressions for the effective material behavior are impossible to evaluate explicitly, in general, such that approximations or bounds for the effective behavior are sought for.

Statistical and material assumptions. In this work, statistical homogeneous and ergodic random materials will be assumed from this point on, see, e.g., Kröner (1977), Willis (1981) or Torquato (2002). Further, the existence of a finite RVE v will be assumed. The ensemble average of a quantity will be considered as the volume average over the RVE, i.e.,

$$\langle q \rangle = \langle q \rangle_v = \frac{1}{v}\int_v q(\boldsymbol{x})\mathrm{d}v\,. \tag{3.28}$$

Under these conditions, the effective stiffness $\bar{\mathbb{C}}$ simplifies to a constant, see Milton (2002), and can be determined by solving the boundary value problem over the RVE loaded on the boundary with a set of six linear independent, effective strains $\bar{\varepsilon}$.

The local elastic potential W, the effective potential $\bar{W}$, the effective strain $\bar{\varepsilon} = \langle \varepsilon \rangle_v$ and the corresponding effective material properties $\bar{\mathbb{C}}$, $\bar{s}$ and $\bar{k}$ of the RVE are defined analogously as

$$\begin{aligned} W &= \tfrac{1}{2}\varepsilon \cdot \mathbb{C}[\varepsilon] - \varepsilon \cdot s - \tfrac{1}{2}k\,, \quad \sigma = \frac{\partial W}{\partial \varepsilon} = \mathbb{C}[\varepsilon] - s\,, \\ \bar{W} &= \langle W \rangle_v = \tfrac{1}{2}(\langle \varepsilon \cdot \sigma - \varepsilon \cdot s - k \rangle_v) = \tfrac{1}{2}\bar{\varepsilon} \cdot \bar{\mathbb{C}}[\bar{\varepsilon}] - \bar{\varepsilon} \cdot \bar{s} - \tfrac{1}{2}\bar{k}\,. \end{aligned} \tag{3.29}$$

In this work, single-phase (SPPM) and multiphase polycrystalline materials (MPPM) are of interest. For an MPPM with m phases, the material properties $\mathbb{C}$, $\boldsymbol{s}$ and k are assumed in the RVE as phasewise constant with local orientations. This implies, e.g., for the stiffness with constant single crystal stiffnesses $\tilde{\mathbb{C}}_p$ of phase p

$$\mathbb{C}(\boldsymbol{x}) = \sum_{p=1}^{m} I^p(\boldsymbol{x})\mathbb{C}_p(\boldsymbol{x}) \ , \quad \mathbb{C}_p(\boldsymbol{x}) = \boldsymbol{Q}_p(\boldsymbol{x}) \star \tilde{\mathbb{C}}_p \tag{3.30}$$

and its average over the RVE

$$\langle\mathbb{C}\rangle_v = \frac{1}{v}\int_v \mathbb{C}(\boldsymbol{x})\mathrm{d}v = \sum_{p=1}^{m} \underbrace{\frac{v_p}{v}}_{=c_p} \underbrace{\frac{1}{v_p}\int_{v_p} \mathbb{C}_p(\boldsymbol{x})\mathrm{d}v}_{=\langle\mathbb{C}_p\rangle_{v_p}} \ . \tag{3.31}$$

The phase average $\langle\mathbb{C}_p\rangle_{v_p}$ can be transformed into an orientation average with the corresponding CODF $f_p(\boldsymbol{Q})$. Analogous treatment is considered for $\boldsymbol{s}$ (with corresponding $\tilde{\boldsymbol{s}}_p$) and k (with corresponding $\tilde{k}_p$), where the orientation averages are considered only for $\boldsymbol{s}$, for k is assumed orientation independent. It is reminded, that each crystallographic orientation of each constituent is to be considered as a distinct phase.

3.4 Bounds of linear properties

3.4.1 General relations

Effective potentials of the RVE. In the RVE volume v, for the convex stress potential $W = W(\boldsymbol{\varepsilon})$ with $\boldsymbol{\sigma} = \partial W/\partial\boldsymbol{\varepsilon}$ given in Eq. (3.29), the effective elastic potential $\bar{W} = \bar{W}(\bar{\boldsymbol{\varepsilon}})$ is defined through the principle of minimum potential energy as

$$\bar{W} = \inf_{\boldsymbol{u}\in K} \langle W\rangle_v \ , \ K = \{\boldsymbol{u} \mid \boldsymbol{u} = \bar{\boldsymbol{\varepsilon}}\boldsymbol{x} \ \forall \boldsymbol{x} \in \partial v\} \ . \tag{3.32}$$

The effective strain is imposed on the boundary of the RVE, which equals the volume average over the RVE, i.e., $\bar{\varepsilon} = \langle \varepsilon \rangle_v$, if the displacement field is continuous, which is assumed in this work.

In Ponte Castañeda and Suquet (1997) (page 184) it is shown that, based on Hills' Lemma, i.e., $\langle \boldsymbol{\sigma} \cdot \varepsilon \rangle_v = \langle \boldsymbol{\sigma} \rangle_v \cdot \langle \varepsilon \rangle_v$ for any divergence free $\boldsymbol{\sigma}$ with continuous traction vector and kinematically compatible ε of a continuous displacement field, the average stress $\langle \boldsymbol{\sigma} \rangle_v$ in linear and nonlinear materials fulfills

$$\langle \boldsymbol{\sigma} \rangle_v = \frac{\partial \bar{W}}{\partial \bar{\varepsilon}} . \tag{3.33}$$

The effective stress $\bar{\boldsymbol{\sigma}}$ of the RVE is defined as $\bar{\boldsymbol{\sigma}} = \langle \boldsymbol{\sigma} \rangle_v$.

The dual problem in terms of the complementary energy density $U = U(\boldsymbol{\sigma}) = W^*(\boldsymbol{\sigma})$, defined through the Legendre-Fenchel transform W^* of W

$$W^* = \sup_{\boldsymbol{\varepsilon}} \{\boldsymbol{\sigma} \cdot \varepsilon - W\} = \frac{1}{2} \boldsymbol{\sigma} \cdot \mathbb{S}[\boldsymbol{\sigma}] + \boldsymbol{\sigma} \cdot \boldsymbol{e} + \frac{1}{2} l \,, \tag{3.34}$$

with the relations

$$\mathbb{S} = \mathbb{C}^{-1} \,, \quad \boldsymbol{e} = \mathbb{C}^{-1}[\boldsymbol{s}] \,, \quad l = k + \boldsymbol{s} \cdot \mathbb{C}^{-1}[\boldsymbol{s}] \,, \tag{3.35}$$

defines the effective complementary potential $\bar{U} = \bar{U}(\bar{\boldsymbol{\sigma}})$ as

$$\begin{gathered} \bar{U} = \inf_{\boldsymbol{\sigma} \in S} \langle U \rangle_v = \frac{1}{2} \bar{\boldsymbol{\sigma}} \cdot \bar{\mathbb{S}}[\bar{\boldsymbol{\sigma}}] + \bar{\boldsymbol{\sigma}} \cdot \bar{\boldsymbol{e}} + \frac{1}{2} \bar{l} \,, \\ S = \{\boldsymbol{\sigma} \mid \operatorname{div}(\boldsymbol{\sigma}) = \boldsymbol{o} \wedge \langle \boldsymbol{\sigma} \rangle_v = \bar{\boldsymbol{\sigma}}\} \,, \end{gathered} \tag{3.36}$$

where $\bar{\boldsymbol{\sigma}}$ is imposed. The average strain fulfills

$$\langle \varepsilon \rangle_v = \frac{\partial \bar{U}}{\partial \bar{\boldsymbol{\sigma}}} . \tag{3.37}$$

In Ponte Castañeda and Suquet (1997) (page 184) it is shown that $\bar{W}$ and $\bar{U}$ (defined as given in Eq. (3.32) and Eq. (3.36)) are Legendre duals, i.e.,

$$\bar{W} + \bar{U} = \bar{\varepsilon} \cdot \bar{\boldsymbol{\sigma}} \,. \tag{3.38}$$

This means that the effective complementary energy density $\bar{U}$ is the solution of the optimization problem

$$\bar{U} = \sup_{\bar{\varepsilon}} \{ \bar{\boldsymbol{\sigma}} \cdot \bar{\varepsilon} - \bar{W} \} \tag{3.39}$$

which is nothing else that the Legendre-Fenchel transform $\bar{W}^* = \bar{U}$ of the effective potential $\bar{W}$.

Bounds. Based on different variational principles, the effective elastic potential $\bar{W}$ might be enclosed by bounds, independently of the loading conditions. Such upper $W^+ = W^+(\bar{\varepsilon})$ and lower $W^- = W^-(\bar{\varepsilon})$ bounds fulfill the inequalitites

$$W^- \leq \bar{W} \leq W^+ \quad \forall \bar{\varepsilon} \,. \tag{3.40}$$

Accessible lower and upper bounds $W^{\pm}$ of $\bar{W}$ deliver also bounds of $\bar{U}$. Based on Eq. (3.39), it follows that $\bar{\boldsymbol{\sigma}} \cdot \bar{\varepsilon} - \bar{W} \geq \bar{\boldsymbol{\sigma}} \cdot \bar{\varepsilon} - W^+ \ \forall \bar{\varepsilon}$ holds. This implies the same ordering for the suprema, such that

$$\underbrace{\sup_{\bar{\varepsilon}} \{ \bar{\boldsymbol{\sigma}} \cdot \bar{\varepsilon} - W^+ \}}_{=(W^+)^* = U^-} \leq \underbrace{\sup_{\bar{\varepsilon}} \{ \bar{\boldsymbol{\sigma}} \cdot \bar{\varepsilon} - \bar{W} \}}_{=\bar{W}^* = \bar{U}} \leq \underbrace{\sup_{\bar{\varepsilon}} \{ \bar{\boldsymbol{\sigma}} \cdot \bar{\varepsilon} - W^- \}}_{=(W^-)^* = U^+} \tag{3.41}$$

holds, i.e., the bounds of $\bar{W}$ can be Legendre-Fenchel transformed in order to obtain bounds for $\bar{U}$.

n-th-order bounds of elastic properties. Consider for a moment purely elastic problems, i.e., $\boldsymbol{s} = \mathbf{0}$ and $k = 0$. The effective elastic energy density is described as $\bar{W} = \frac{1}{2} \bar{\varepsilon} \cdot \bar{\mathbb{C}}[\bar{\varepsilon}]$. Based on n-point probability information of the RVE, n-th-order bounds can be constructed, see, e.g., Kröner (1977) and Willis (1981). High-order bounds give access to a better understanding of the space in which the effective properties are assured to be contained but require a high amount of computation, especially if the full infinite-dimensional function space of n-point-statistics (n-PPFs) is to be evaluated. Low-order bounds give naturally only rough information of the effective properties but might contain crucial information for material design purposes. The effective stiffness $\bar{\mathbb{C}}$ of the RVE with region v is said to be bounded from below and above by the lower / upper n-th-order bounds $\mathbb{C}^{n\pm}$ (which take into consideration n-point probability information of the RVE) if

$$\bar{\varepsilon} \cdot \mathbb{C}^{n-}[\bar{\varepsilon}] \leq \bar{\varepsilon} \cdot \bar{\mathbb{C}}[\bar{\varepsilon}] \leq \bar{\varepsilon} \cdot \mathbb{C}^{n+}[\bar{\varepsilon}] \quad \forall \bar{\varepsilon} \tag{3.42}$$

hold. This kind of inequalities are naturally connected to the corresponding effective potential bounds $W^{n\pm}$ and will be addressed shortly from now on as $\mathbb{C}^{n-} \leq \bar{\mathbb{C}} \leq \mathbb{C}^{n+}$. It should be remarked that, based on matrix analysis, see, e.g., Horn and Johnson (1990), if $\mathbb{C}^{n-} \leq \bar{\mathbb{C}} \leq \mathbb{C}^{n+}$ holds, then $(\mathbb{C}^{n+})^{-1} \leq \bar{\mathbb{C}}^{-1} \leq (\mathbb{C}^{n-})^{-1}$ also holds. Thus, the lower stiffness bound might also be considered as the upper compliance bound. The same results can be obtained based on Eq. (3.41).
From matrix analysis, it is known, that if Eq. (3.42) holds, then

$$\lambda_\alpha^{n-} \leq \bar{\lambda}_\alpha \leq \lambda_\alpha^{n+}, \quad \alpha = 1, ..., \beta \tag{3.43}$$

necessarily holds for the β corresponding eigenvalues, where each of the sets $\{\lambda_\alpha^{n-}\}, \{\bar{\lambda}_\alpha\}, \{\lambda_\alpha^{n+}\}$ is sorted in increasing order. From these necessary conditions, other necessary conditions are easily derivable, e.g., for the trace, determinant, Frobenius norm or any invariant of the considered tensors viewed as linear maps for the space of symmetric second-order tensors.

For given bounds $\mathbb{C}^{n\pm}$ (of arbitrary order), based on Eq. (3.42), it is possible to give explicit bounds not only on the "diagonal" components of the enclosed $\bar{\mathbb{C}}$, but also on the "off-diagonal" components of the effective stiffness $\bar{\mathbb{C}}$

$$\begin{aligned} \Gamma^{n-}_{\alpha\beta} \leq \bar{C}_{\alpha\beta} \leq \Gamma^{n+}_{\alpha\beta} \quad \alpha, \beta \in \{11, 22, 33, 23, 13, 12\} \\ \Gamma^{n\pm}_{\alpha\beta} = \tfrac{1}{2}(C^{n+}_{\alpha\beta} + C^{n-}_{\alpha\beta}) \pm \tfrac{1}{2}\sqrt{(C^{n+}_{\alpha\alpha} - C^{n-}_{\alpha\alpha})(C^{n+}_{\beta\beta} - C^{n-}_{\beta\beta})}\,, \end{aligned} \tag{3.44}$$

see Lobos and Böhlke (2016) or Appendix C of this work. These components bounds improve the ones given in Proust and Kalidindi (2006) for major symmetric $\bar{\mathbb{C}}$ and can be used by materials designers iteratively if more and more statistical information of the considered inhomogeneous material is available.

Also, it should be remarked that, based on Eq. (3.42), mechanical quantities defined as $q = \boldsymbol{A} \cdot \mathbb{B}[\boldsymbol{A}]$ (and variations of the quadratic form of a fourth-order tensor $\mathbb{B}$) will obey the bounds hierarchy. This means, e.g., that for Young's modulus E of a material with stiffness $\mathbb{C}$ in tensile direction $\boldsymbol{n}$ defined as $E(\mathbb{C}, \boldsymbol{n}) = 1/(\boldsymbol{n}^{\otimes 2} \cdot \mathbb{C}^{-1}[\boldsymbol{n}^{\otimes 2}])$,

$$E^{n-} \leq \bar{E} \leq E^{n+}\,, \quad E^{n\pm} = E(\mathbb{C}^{n\pm}, \boldsymbol{n})\,, \quad \bar{E} = E(\bar{\mathbb{C}}, \boldsymbol{n}) \tag{3.45}$$

holds for given tensile direction $\boldsymbol{n}$. For mechanical quantities defined as $q = \boldsymbol{A} \cdot \mathbb{B}[\boldsymbol{C}]$, based only on Eq. (3.42), no statement can be made, in general.

Finally, in this work, motivated by Lobos and Böhlke (2016), and the derivable bounds for the determinant, we define the relative elastic volume

$$e_n = \frac{\det(\mathbb{C}^{n+})}{\det(\mathbb{C}^{n-})} \in [1, \infty)\,. \tag{3.46}$$

The relative elastic volume, due to the properties of the determinant, is invariant under consideration of the inverse bounds for the effective compliance. The determinant takes into account the multiplicity of the different eigenvalues and is, therefore, assumed to be an appropriate description for the size of the elasticity space spanned by corresponding bounds.

Extension to eigenfields. Consider now again the linear problem with eigenfields $\boldsymbol{s} \neq \mathbf{0}$ and $k \neq 0$. We introduce the scalar quantity $\bar{W}_{\underline{\varepsilon}}$

$$\bar{W}_{\underline{\varepsilon}} = \frac{1}{2}\underline{\varepsilon}^{\mathsf{T}}\,\underline{\underline{\bar{C}}}\,\underline{\varepsilon} \tag{3.47}$$

with the supervector $\underline{\varepsilon} = [\bar{\boldsymbol{\varepsilon}}, y]^{\mathsf{T}}$ (with real-valued y) and the supertensor

$$\underline{\underline{\bar{C}}} = \begin{bmatrix} \bar{\mathbb{C}} & -\bar{\boldsymbol{s}} \\ -\bar{\boldsymbol{s}} & -\bar{k} \end{bmatrix}, \quad \underline{\underline{\bar{S}}} = \underline{\underline{\bar{C}}}^{*} = \begin{bmatrix} \bar{\mathbb{C}}^{-1} & \bar{\mathbb{C}}^{-1}[\bar{\boldsymbol{s}}] \\ \bar{\mathbb{C}}^{-1}[\bar{\boldsymbol{s}}] & \bar{k} + \bar{\boldsymbol{s}} \cdot \bar{\mathbb{C}}^{-1}[\bar{\boldsymbol{s}}] \end{bmatrix} \tag{3.48}$$

which might be considered as a corresponding symmetric 7×7 matrices for linear elasticity with eigenfields. The symbol $\underline{\underline{\bar{C}}}^{*}$ represents the supertensor in the complementary space obtained through the Legendre-Fenchel transformation of $\bar{W}$, see analogously Eq. (3.34) and Eq. (3.35). For a supertensor in the complementary space $\underline{\underline{\bar{S}}}$ we define

$$\underline{\underline{\bar{S}}} = \begin{bmatrix} \bar{\mathbb{S}} & \bar{\boldsymbol{e}} \\ \bar{\boldsymbol{e}} & \bar{l} \end{bmatrix}, \quad \underline{\underline{\bar{S}}}^{*} = \begin{bmatrix} \bar{\mathbb{S}}^{-1} & -\bar{\mathbb{S}}^{-1}[\bar{\boldsymbol{e}}] \\ -\bar{\mathbb{S}}^{-1}[\bar{\boldsymbol{e}}] & -(\bar{l} - \bar{\boldsymbol{e}} \cdot \bar{\mathbb{S}}^{-1}[\bar{\boldsymbol{e}}]) \end{bmatrix}. \tag{3.49}$$

It follows that $(\underline{\underline{\bar{C}}}^{*})^{*} = \underline{\underline{\bar{C}}}$ holds for the expressions of this work.
For $\underline{\varepsilon} = [\bar{\boldsymbol{\varepsilon}}, 1]^{\mathsf{T}}$, clearly $\bar{W}_{\underline{\varepsilon}} = \bar{W}$ holds. If lower and upper bounds $W^{\pm}$ of $\bar{W}$, represented as

$$W^{\pm} = \frac{1}{2}\bar{\boldsymbol{\varepsilon}} \cdot \mathbb{C}^{\pm}[\bar{\boldsymbol{\varepsilon}}] - \bar{\boldsymbol{\varepsilon}} \cdot \boldsymbol{s}^{\pm} - \frac{1}{2}k^{\pm}, \tag{3.50}$$

are obtainable, then it can be shown based on Appendix C that the corresponding supertensors

$$\underline{\underline{C}}^{\pm} = \begin{bmatrix} \mathbb{C}^{\pm} & -\boldsymbol{s}^{\pm} \\ -\boldsymbol{s}^{\pm} & -k^{\pm} \end{bmatrix} \tag{3.51}$$

fulfill

$$\underline{\varepsilon}^{\mathsf{T}}\, \underline{\underline{C}}^{-}\, \underline{\varepsilon} \leq \underline{\varepsilon}^{\mathsf{T}}\, \underline{\underline{\bar{C}}}\, \underline{\varepsilon} \leq \underline{\varepsilon}^{\mathsf{T}}\, \underline{\underline{C}}^{+}\, \underline{\varepsilon} \quad \forall \underline{\varepsilon}\,, \tag{3.52}$$

abbreviated as $\underline{\underline{C}}^{-} \leq \underline{\underline{\bar{C}}} \leq \underline{\underline{C}}^{+}$. It is then clear that, based on the component bounds given in Lobos and Böhlke (2016) for off-diagonal matrix components, see, e.g., Appendix C, the effective $\bar{\boldsymbol{s}}$ and $\bar{k}$ can be bounded based on the values in the corresponding rows and columns of $\underline{\underline{C}}^{\pm}$. It should be stressed, that this naturally holds not only for linear elasticity but for every other analogous physical problem based on elliptic differential equations with corresponding physical eigenfields.

Explicitly, for linear elasticity with eigenfields, ε is a symmetric tensor for which

$$\begin{aligned} \boldsymbol{B}_1 &= \boldsymbol{b}_{11}\,, & \boldsymbol{B}_4 &= \tfrac{1}{\sqrt{2}}(\boldsymbol{b}_{23} + \boldsymbol{b}_{32})\,, \\ \boldsymbol{B}_2 &= \boldsymbol{b}_{22}\,, & \boldsymbol{B}_5 &= \tfrac{1}{\sqrt{2}}(\boldsymbol{b}_{13} + \boldsymbol{b}_{31})\,, \\ \boldsymbol{B}_3 &= \boldsymbol{b}_{33}\,, & \boldsymbol{B}_6 &= \tfrac{1}{\sqrt{2}}(\boldsymbol{b}_{12} + \boldsymbol{b}_{21})\,, \end{aligned} \tag{3.53}$$

constitute an orthonormal basis for symmetric tensors, see Appendix A. Every symmetric tensor $\boldsymbol{A} = \sum_{i,j=1}^{3} A_{ij}\boldsymbol{b}_{ij}$ is then represented as a vector as

$$\boldsymbol{A} = \sum_{i=1}^{6} A_i \boldsymbol{B}_i\,, \quad A_i = \begin{bmatrix} A_{11} & A_{22} & A_{33} & \sqrt{2}A_{23} & \sqrt{2}A_{13} & \sqrt{2}A_{12} \end{bmatrix}^{\mathsf{T}}. \tag{3.54}$$

Defining an artificial seventh base tensor $\boldsymbol{B}_7$, with which $\{\boldsymbol{B}_i\}$ constitute an orthonormal basis, allows representing the supervector $\underline{\varepsilon} = [\bar{\varepsilon}, y]^{\mathsf{T}}$ and the supertensor $\underline{\underline{\bar{C}}}$ containing the material properties with vector and matrix components as

$$\underline{\varepsilon} = \sum_{i=1}^{7} (\underline{\varepsilon})_i \boldsymbol{B}_i \, , \quad \underline{\underline{\bar{C}}} = \sum_{i,j=1}^{7} (\underline{\underline{\bar{C}}})_{ij} \boldsymbol{B}_i \otimes \boldsymbol{B}_j \tag{3.55}$$

with

$$(\underline{\varepsilon})_i = \begin{bmatrix} \bar{\varepsilon}_{11} & \bar{\varepsilon}_{22} & \bar{\varepsilon}_{33} & \sqrt{2}\bar{\varepsilon}_{23} & \sqrt{2}\bar{\varepsilon}_{13} & \sqrt{2}\bar{\varepsilon}_{12} & y \end{bmatrix}^{\mathsf{T}} \tag{3.56}$$

and

$$(\underline{\underline{\bar{C}}})_{ij} = \begin{bmatrix} \bar{C}_{1111} & \bar{C}_{1122} & \bar{C}_{1133} & \sqrt{2}\bar{C}_{1123} & \sqrt{2}\bar{C}_{1113} & \sqrt{2}\bar{C}_{1112} & -\bar{s}_{11} \\ \bar{C}_{1122} & \bar{C}_{2222} & \bar{C}_{2233} & \sqrt{2}\bar{C}_{2223} & \sqrt{2}\bar{C}_{1322} & \sqrt{2}\bar{C}_{1222} & -\bar{s}_{22} \\ \bar{C}_{1133} & \bar{C}_{2233} & \bar{C}_{3333} & \sqrt{2}\bar{C}_{2333} & \sqrt{2}\bar{C}_{1333} & \sqrt{2}\bar{C}_{1233} & -\bar{s}_{33} \\ \sqrt{2}\bar{C}_{1123} & \sqrt{2}\bar{C}_{2223} & \sqrt{2}\bar{C}_{2333} & 2\bar{C}_{2323} & 2\bar{C}_{1323} & 2\bar{C}_{1223} & -\sqrt{2}\bar{s}_{23} \\ \sqrt{2}\bar{C}_{1113} & \sqrt{2}\bar{C}_{1322} & \sqrt{2}\bar{C}_{1333} & 2\bar{C}_{1323} & 2\bar{C}_{1313} & 2\bar{C}_{1213} & -\sqrt{2}\bar{s}_{13} \\ \sqrt{2}\bar{C}_{1112} & \sqrt{2}\bar{C}_{1222} & \sqrt{2}\bar{C}_{1233} & 2\bar{C}_{1223} & 2\bar{C}_{1213} & 2\bar{C}_{1212} & -\sqrt{2}\bar{s}_{12} \\ -\bar{s}_{11} & -\bar{s}_{22} & -\bar{s}_{33} & -\sqrt{2}\bar{s}_{23} & -\sqrt{2}\bar{s}_{13} & -\sqrt{2}\bar{s}_{12} & -\bar{k} \end{bmatrix} . \tag{3.57}$$

The given vectors and matrices are in accordance with the expected state variable $\underline{\sigma} = [\bar{\boldsymbol{\sigma}}, z] = \underline{\underline{\bar{C}}}\,\underline{\varepsilon}$ (with corresponding real $z = -\bar{\boldsymbol{\varepsilon}} \cdot \bar{\boldsymbol{s}} - \bar{k}y$) and expected quadratic form $\bar{W}_{\underline{\varepsilon}} = \frac{1}{2}\bar{\varepsilon}\bar{\mathbb{C}}[\bar{\varepsilon}] - \bar{\varepsilon} \cdot \bar{\boldsymbol{s}}y - \frac{1}{2}\bar{k}y^2$. Based on the matrix Eq. (3.57), and corresponding matrices for $\underline{\underline{C}}^{\pm}$, e.g., the effective eigenfield component $-\bar{s}_{33}$ (i.e., the $(3,7)$ component in the matrix) is bounded analogously as in Eq. (3.44) as

$$\begin{aligned} \gamma_{37}^{-} \leq -\bar{s}_{33} \leq \gamma_{37}^{+} \, , \quad \gamma_{37}^{\pm} &= \mu_{37} \pm \Delta_{37} \, , \\ \mu_{37} &= \tfrac{1}{2}(-s_{33}^{+} + (-1)s_{33}^{-}) \, , \\ \Delta_{37} &= \tfrac{1}{2}\sqrt{(C_{3333}^{+} - C_{3333}^{-})(-k^{+} - (-1)k^{-})} \, . \end{aligned} \tag{3.58}$$

One extraordinary property is visible in Eq. (3.58), namely, for the case $k^{+} = k^{-}$, the exact $\bar{\boldsymbol{s}}$ is obtained immediately as $\bar{\boldsymbol{s}} = \boldsymbol{s}^{+} = \boldsymbol{s}^{-}$. From a micromechanical point of view, if the bounds $W^{\pm}$ depend on statistical quantities of the composite, and microstructures of a material can be found for which $k^+ = k^-$ holds, then $\bar{s}$ is obtained immediately without the need for any further computation. The search for materials for which such microstructures exist is an intriguing quest for materials research and design. Even if no real world material exists for which $k^+ = k^-$ holds for any derivable bounds, the term $(k^+ - k^-)^2$ might be minimized in order to make $\bar{s}$ more controllable.

3.4.2 Zeroth-order bounds

Zeroth-order bounds of linear elastic properties. It is true, that based on n-point information of the RVE with $n \geq 1$, n-th-order bounds reflect important information about the heterogeneous material. But it is also important to realize that the effective properties might also be bounded solely by the m material constituents of the heterogeneous material, *independently* of the microstructural arrangement. These are the zeroth-order bounds $\mathbb{C}^{0\pm}$, introduced by Kröner (1977), investigated by Nadeau and Ferrari (2001) and improved by Lobos and Böhlke (2016). The zeroth-order bounds are isotropic tensors $\mathbb{C}^I = \lambda_1 \mathbb{P}_1 + \lambda_2 \mathbb{P}_2 \geq \mathbb{C}(\boldsymbol{x}) \forall \boldsymbol{x} \in v$, as discussed by Kröner (1977) and shown in Nadeau and Ferrari (2001). For an MPPM the condition $\mathbb{C}^I \geq \mathbb{C}(\boldsymbol{x}) \forall \boldsymbol{x} \in v$ is equivalent to $\mathbb{C}^I \geq \tilde{\mathbb{C}}_p \forall p$, where $\tilde{\mathbb{C}}_p$ denotes the single crystal stiffness of phase p. The upper zeroth-order bound $\mathbb{C}^{0+} = \sum_{\alpha=1}^{2} \lambda_\alpha^{0+} \mathbb{P}_\alpha$ is defined in Lobos and Böhlke (2016) following Nadeau and Ferrari (2001) through the optimization problem

$$
\begin{aligned}
\{\lambda_1^{0+}, \lambda_2^{0+}\} \quad &= \quad \arg \min_{\{\lambda_\alpha\}} \operatorname{tr}\left(\lambda_1 \mathbb{P}_1 + \lambda_2 \mathbb{P}_2\right) \\
&\quad \text{such that} \quad \tilde{\mathbb{C}}_p \leq \left(\lambda_1 \mathbb{P}_1 + \lambda_2 \mathbb{P}_2\right) \forall p \,.
\end{aligned}
\tag{3.59}
$$

For an SPPM, the solution of this optimization problem is easily obtained with the elegant decompositions presented in Nadeau and Ferrari (2001), which delivered the following optimization problem for the trace of the stiffness tensor

$$\begin{aligned} \beta(\alpha) &= \{\text{largest eigenvalue of } \tilde{\mathbb{C}}_p - \alpha\mathbb{P}_1\} \\ \alpha^+ &= \arg\min_{\alpha}\ \mathrm{tr}\left(\alpha\mathbb{P}_1 + \beta(\alpha)\mathbb{I}^S\right) \\ \{\lambda^{0+}_{p,1}, \lambda^{0+}_{p,2}\} &= \{\alpha^+ + \beta(\alpha+), \beta(\alpha^+)\}\,. \end{aligned} \tag{3.60}$$

This delivers the upper zeroth-order bound $\mathbb{C}^{0+}_p = \sum_{\alpha=1}^{2} \lambda^{0+}_{p,\alpha}\mathbb{P}_\alpha$ of phase p.

The upper zeroth-order bound $\mathbb{C}^{0+}$ of an MPPM defined through Eq. (3.59) is more difficult to compute, in general. One way to obtain an admissible upper bound fulfilling $\tilde{\mathbb{C}}_p \leq \tilde{\mathbb{C}}^{0+} \forall p$ is to compute all $\mathbb{C}^{0+}_p$ separately and setting

$$\tilde{\mathbb{C}}^{0+} = \tilde{\lambda}^{0+}_1\mathbb{P}_1 + \tilde{\lambda}^{0+}_2\mathbb{P}_2\,,\ \tilde{\lambda}^{0+}_1 = \max_p\{\lambda^{0+}_{p,1}\}\,,\ \tilde{\lambda}^{0+}_2 = \max_p\{\lambda^{0+}_{p,2}\}\,. \tag{3.61}$$

However, $\tilde{\mathbb{C}}^{0+}$ has no minimal trace, in general, in the set described by $\tilde{\mathbb{C}}_p \leq \mathbb{C}^I\ \forall p$, see Appendix C, and the example described from Eq. (C.6) to Eq. (C.10). The solution of Eq. (3.59) can be computed based on algorithms for positive semidefinite programming with square slack variables, see Appendix C. From the perspective of material databases, the zeroth-order bound $\tilde{\mathbb{C}}^{0+}$ is far more attractive than the solution $\mathbb{C}^{0+}$ of Eq. (3.59). For a given material database, the zeroth-order bound of each material can be computed a priori. Then, for arbitrary combination of these materials, evaluation of $\tilde{\mathbb{C}}^{0+}$ is trivial, while for every new combination of materials $\mathbb{C}^{0+}$ needs to be computed again and again. For small material databases, $\mathbb{C}^{0+}$ might be achievable, but for large databases, it is recommended to use the admissible bound $\tilde{\mathbb{C}}^{0+}$.

The upper zeroth-order bound $\mathbb{C}^{0+}$ is defined as the isotropic tensor with minimal trace (in the sense of invariants of linear maps) delivering a greater elastic energy density at any point of the RVE. In Lobos and Böhlke (2016) it is discussed that the zeroth-order bounds are not unique and that the definition Eq. (3.59) is only one possibility. The zeroth-order bounds could alternatively be defined on other invariants of the fourth-order tensor. The definition proposed by Kröner (1977) is not valid for arbitrary anisotropic materials, as shown in Lobos et al. (2017) by an explicit example. The lower zeroth-order bound $\mathbb{C}^{0-}$ is defined in Lobos and Böhlke (2016) as the upper zeroth-order bound of the compliance, i.e., $\bar{\mathbb{C}}^{-1} \leq (\mathbb{C}^{0-})^{-1}$, and solving the corresponding optimization problem analogous to Eq. (3.59). This definition ensures the positive definiteness of both zeroth-order bounds, which is not the case as for the definitions proposed in Nadeau and Ferrari (2001), see Lobos and Böhlke (2016) for a discussion.

Anisotropy measure of SPPMs and MPPMs. The scalar e_0 based on Eq. (3.46) and the zeroth-order bounds, see Lobos and Böhlke (2016), is able to reflect several properties of materials. The zeroth-order relative elastic volume e_0 for single-phase materials reflects the intrinsic anisotropy of the SPPM itself. For isotropic and anisotropic materials, $e_0 = 1$ and $e_0 > 1$ hold, respectively. Without any modifications, e_0 is also able to describe the intrinsic anisotropy strength of MPPMs. For example, if a multiphase material with two anisotropic phases is considered, then $e_0 > 1$ might be expected. But, the two anisotropic stiffness tensors of the considered phases might have both high values not varying much from each other. For such an MPPM, e_0 will not be large. But, e.g., another 2-phase material with isotropic phases having large relative distances will be able to show highly anisotropic material behavior. Those multiphase materials are characterized by large values of e_0 reflecting the large relative space of possibilities for material designers in different applications.

Naturally, if information about the specific range of the different components or other mechanical quantities is desired, then instead of e_0 the corresponding bounds, given in Eq. (3.44) and Eq. (3.45), can be examined.

Bounds for local elastic energy density. The zeroth-order bounds, compared to all other upcoming bounds, have the property not only to bound the effective energy density $\bar{W}$, but also the local energy density $W(\boldsymbol{x})$ at every point of the material. If a material characteristic critical energy level $W_{p,c}$ of phase p is not to be surpassed at a material point in the material region v_p of phase p, then the following estimate for $\boldsymbol{x} \in v_p$ can be considered

$$\begin{aligned} 2W(\boldsymbol{x}) &= \boldsymbol{\varepsilon}(\boldsymbol{x}) \cdot \mathbb{C}(\boldsymbol{x})[\boldsymbol{\varepsilon}(\boldsymbol{x})] \\ &\leq \boldsymbol{\varepsilon}(\boldsymbol{x}) \cdot \mathbb{C}_p^{0+}[\boldsymbol{\varepsilon}(\boldsymbol{x})] \\ &= (\lambda_{p,1}^{0+} - \lambda_{p,2}^{0+}) \tfrac{\operatorname{tr}(\boldsymbol{\varepsilon}(\boldsymbol{x}))^2}{3} + \lambda_{p,2}^{0+} \|\boldsymbol{\varepsilon}(\boldsymbol{x})\|^2 \\ &\leq \tfrac{1}{3}(\lambda_{p,1}^{0+} + 2\lambda_{p,2}^{0+}) \|\boldsymbol{\varepsilon}(\boldsymbol{x})\|^2 \leq 2W_{p,c} \,. \end{aligned} \tag{3.62}$$

This estimate motivates the definition of the critical strain magnitude

$$\varepsilon_{p,c} = \sqrt{\frac{6W_{p,c}}{\lambda_{p,1}^{0+} + 2\lambda_{p,2}^{0+}}} \,, \tag{3.63}$$

which depends only on material characteristics of the phase. This result offers two perspectives for applications. Single-phase materials might be characterized by this critical strain value, or, materials being able to bear a prescribed maximum strain $\max_{\boldsymbol{x} \in v_p} \|\boldsymbol{\varepsilon}(\boldsymbol{x})\| \leq \varepsilon_{p,c}$ can be selected from given material data. Naturally, these critical strain measures are not valid for phase boundaries but might be useful for single-phase materials.

It should also be noted that analogous dual formulations for the complementary energy density, stress and zeroth-order bounds for the compliance are obtainable.

Extension to eigenfields. Due to the definition of $\bar{W}$ through the principle of minimum potential energy

$$\bar{W} \leq \langle W(\check{\varepsilon})\rangle_v \quad \forall \check{\boldsymbol{u}} \in K\ , \quad \check{\varepsilon} = \mathrm{sym}(\mathrm{grad}(\check{\boldsymbol{u}})) \tag{3.64}$$

upper bounds of $\bar{W}$ might be obtained by estimating $\langle W(\check{\varepsilon})\rangle_v$. By using the notation introduced in Eq. (3.47), $W(\check{\varepsilon})$ can be equally described as

$$W_{\underline{\check{\varepsilon}}} = \frac{1}{2}\underline{\check{\varepsilon}}^{\mathsf{T}}\, \underline{\underline{C}}\, \underline{\check{\varepsilon}}\,, \quad \underline{\underline{C}} = \begin{bmatrix} \mathbb{C} & -\boldsymbol{s} \\ -\boldsymbol{s} & -k \end{bmatrix} \tag{3.65}$$

for $\underline{\check{\varepsilon}} = [\check{\varepsilon}, 1]^{\mathsf{T}}$. A orientation $\boldsymbol{Q}$ acts as follows

$$\boldsymbol{Q} \star \underline{\underline{C}} = \begin{bmatrix} \boldsymbol{Q} \star \mathbb{C} & \boldsymbol{Q} \star (-\boldsymbol{s}) \\ \boldsymbol{Q} \star (-\boldsymbol{s}) & \boldsymbol{Q} \star (-k) \end{bmatrix} . \tag{3.66}$$

We now define the constant supertensors

$$\underline{\underline{C}}_0 = \begin{bmatrix} \mathbb{C}_0 & -\boldsymbol{s}_0 \\ -\boldsymbol{s}_0 & -k_0 \end{bmatrix} , \quad \underline{\underline{C}} \leq \underline{\underline{C}}_0 \quad \forall \boldsymbol{x} \in v\ . \tag{3.67}$$

For MPPMs, the supertensors $\underline{\underline{C}}_0$ necessarily have to be invariant with respect to any $\boldsymbol{Q}$, i.e., they are isotropic and have the form

$$\underline{\underline{C}}_0 = \begin{bmatrix} (\lambda_{01}\mathbb{P}_1 + \lambda_{02}\mathbb{P}_2) & -s_0\boldsymbol{I} \\ -s_0\boldsymbol{I} & -k_0 \end{bmatrix} . \tag{3.68}$$

As for the zeroth-order bounds of elasticity, no unique zeroth-order bound exist. In the present work it is proposed to define the general upper zeroth-order bound

$$\underline{\underline{C}}^{0+} = \begin{bmatrix} (\lambda_1^{0+}\mathbb{P}_1 + \lambda_2^{0+}\mathbb{P}_2) & -s^{0+}\boldsymbol{I} \\ -s^{0+}\boldsymbol{I} & -k^{0+} \end{bmatrix} \tag{3.69}$$

following Lobos and Böhlke (2016) with $\operatorname{tr}(\underline{\underline{C}}) = \operatorname{tr}(\mathbb{C}) - k$ by the optimization problem

$$
\begin{aligned}
\{\lambda_1^{0+}, \lambda_2^{0+}, s^{0+}, k^{0+}\} &= \arg \min_{\{\lambda_{01}, \lambda_{02}, s_0, k_0\}} \operatorname{tr}\left(\underline{\underline{C}}_0\right) \\
&\quad \text{such that} \quad \underline{\underline{C}}_p \leq \underline{\underline{C}}_0 \ \forall p ,
\end{aligned}
\tag{3.70}
$$

with the corresponding single crystal properties $\underline{\underline{C}}_p$ of phase p. The solution of this problem might be obtained numerically with algorithms for positive semidefinite programming, as illustrated in Appendix C. The constructed supertensor $\underline{\underline{C}}^{0+}$ fulfills $\underline{\underline{C}} \leq \underline{\underline{C}}^{0+}$ for all $\underline{\check{\varepsilon}}$ (also for $\underline{\check{\varepsilon}} = [\check{\varepsilon}, 1]^{\mathsf{T}}$) at every $\boldsymbol{x} \in v$. This implies for $\check{\boldsymbol{u}} = \bar{\varepsilon}\boldsymbol{x}$ and $\underline{\check{\varepsilon}} = [\bar{\varepsilon}, 1]^{\mathsf{T}}$

$$
\bar{W} \leq \langle W(\bar{\varepsilon}) \rangle_v = \frac{1}{2} \underline{\check{\varepsilon}} \, \langle \underline{\underline{C}} \rangle_v \underline{\check{\varepsilon}} \leq \frac{1}{2} \underline{\check{\varepsilon}} \, \underline{\underline{C}}^{0+} \underline{\check{\varepsilon}} \quad \forall \bar{\varepsilon} . \tag{3.71}
$$

The lower bound $\underline{\underline{C}}^{0-}$ is analogously defined through the dual problem in complementary space, i.e., $\underline{\underline{S}}^{0+} = (\underline{\underline{C}}^{0-})^*$ with $\underline{\underline{S}} = \underline{\underline{C}}^*$.

Simplification for large material databases. Due to the structure of the supertensors Eq. (3.69), the corresponding eigentensors (and eigenspaces) are not constant. This complicates computations for MPPMs in large material databases, since, as discussed for Eq. (3.61), the zeroth-order bound Eq. (3.70) would have to be computed for every new material combination. This complication can be avoided with the following alternative zeroth-order bounds. First, for an SPPM with corresponding $\underline{\underline{C}}_p$, the admissible zeroth-order bound

$$
\begin{aligned}
\underline{\underline{\tilde{C}}}_p^{0+} &= \begin{bmatrix} (\tilde{\lambda}_{p1}^{0+} \mathbb{P}_1 + \tilde{\lambda}_{p2}^{0+} \mathbb{P}_2) & \mathbf{0} \\ \mathbf{0} & -\tilde{k}_p^{0+} \end{bmatrix} , \\
\{\tilde{\lambda}_{p1}^{0+}, \tilde{\lambda}_{p2}^{0+}, \tilde{k}_p^{0+}\} &= \arg \min_{\{\lambda_{01}, \lambda_{02}, k_0\}} \operatorname{tr}\left(\underline{\underline{C}}_0\right) |_{s_0 = 0} \\
&\quad \text{such that} \quad \underline{\underline{C}}_p \leq \underline{\underline{C}}_0 |_{s_0 = 0} ,
\end{aligned}
\tag{3.72}
$$

can be computed. The bound Eq. (3.72) can be calculated for each material in a material database. The eigenspaces of these alternative zeroth-order bounds are constant such that the admissible zeroth-order bound for an MPPM

$$\begin{aligned} \underline{\underline{\tilde{C}}}^{0+} &= \begin{bmatrix} (\tilde{\lambda}_1^{0+}\mathbb{P}_1 + \tilde{\lambda}_2^{0+}\mathbb{P}_2) & \mathbf{0} \\ \mathbf{0} & -\tilde{k}^{0+} \end{bmatrix} , \\ \tilde{\lambda}_1^{0+} = \max_p\{\tilde{\lambda}_{p1}^{0+}\} \, , \; \tilde{\lambda}_2^{0+} &= \max_p\{\tilde{\lambda}_{p2}^{0+}\} \, , \; -\tilde{k}^{0+} = \max_p\{-\tilde{k}_p^{0+}\} \end{aligned} \tag{3.73}$$

can be evaluated without any further expensive computations and immediately fulfills $\underline{\underline{C}}_p \leq \underline{\underline{\tilde{C}}}^{0+}$. The bound $\underline{\underline{\tilde{C}}}^{0+}$ differs, in general, from the bound $\underline{\underline{C}}^{0+}$.

Benefits for materials design. From a homogenization point of view, the zeroth-order bounds or any combination of them are naturally not to be used as approximations for the effective stiffness. These bounds do not incorporate any information of the microstructure statistics. But from a materials design perspective, exactly this property is of crucial importance and is one contribution of the present work for the field of materials design, as discussed in Lobos and Böhlke (2016). The zeroth-order bounds enclose the effective material behavior *for all* realizations / microstructures of the considered inhomogeneous material, only based on its material phases. This means that, based only on the choice of materials and, e.g., the explicit bounds given in Eq. (3.44) and Eq. (3.45), the material limitations of any realization of the chosen material combination are immediately accessible, without the need for experiments or simulations accounting for the infinite number of possible microstructures. This illustrates the zeroth-order bounds as perfect decision support quantities for prescribed properties-profiles in materials screening. For example, if for a future application, certain material properties are needed, then based on material databases, see, e.g., `https://materialsproject.org/` and de Jong et al. (2015), the zeroth-order bounds of each material (and any number of material combinations) can be computed. Material designers are then immediately able to see which materials can be excluded. This enables the automation of the process of materials selection if only physical properties are considered. For a more efficient evaluation of large material databases, the results for each single-phase material can be stored in the database itself. Based on $\tilde{\mathbb{C}}^{0\pm}$ and $\tilde{\underline{\underline{C}}}^{0\pm}$, evaluation for multiphase materials is simplified significantly.

Naturally, the set of criteria for a sensible materials selection for practical applications is much larger, see Ashby and Johnson (2013). In the present work, in Chapter 4 the zeroth-order bounds will be used in order to select material from example material databases.

3.4.3 First-order bounds

First-order bounds of linear elastic properties. If one-point statistical information (i.e., volume fractions) of the microstructure of the inhomogeneous material is available (or taken into consideration), then the simplest possible computations are the volume average of the stiffness and compliance, see Voigt (1910) and Reuss (1929). In Hill (1952) it is shown that these expressions are the first-order bounds of the effective properties

$$\mathbb{C}^{1+} = \langle \mathbb{C} \rangle_v \,, \quad \mathbb{C}^{1-} = (\langle \mathbb{C}^{-1} \rangle_v)^{-1} \,. \tag{3.74}$$

It should be noted that for arbitrarily anisotropic MPPM, the expression in Eq. (3.74) average only minor and major symmetric fourth-order tensors. This immediately implies that these expressions depend only on the second- and fourth-order texture coefficients of the corresponding CODFs of each phase.

Extension. Actually, already in Eq. (3.71) the extended first-order bound $\underline{\underline{C}}^{1+} = \langle \underline{\underline{C}} \rangle_v$ has been shown. The lower bound can be obtained through the dual problem, $\underline{\underline{C}}^{0-} = (\underline{\underline{S}}^{1+})^*$.

It should be noted that the first-order bounds already offer search options for microstructure statistics fulfilling $k^{1+} = k^{1-}$ or minimizing $(k^{1+} - k^{1-})^2$, as motivated by Eq. (3.58).

Characterization and accuracy. If the material phases and the one-point statistics of a multiphase material are given, then either component bounds (Eq. (3.44)), bounds for mechanical characteristics (Eq. (3.45)), or the relative elastic volume (Eq. (3.46)) might be examined. These quantities can reflect qualitative information of the limitations of higher-order variations. For example, consider given materials and given one-point statistics inducing a small relative elastic volume. For such a case, a combination of the first-order bounds might be a good approximation for the effective stiffness since for the fixed one-point statistics the first-order bounds seem close to each other based on the small relative elastic volume. Large relative elastic volumes reflect that even for fixed one-point statistics, variation of two- and higher-point statistics might cover a large number of different properties.

Variation of one-point statistics and properties-closure. For given materials, for each point in the space of one-point statistics the first-order bounds enclose the effective properties and derivable quantities, see Eq. (3.44) - Eq. (3.46). For example, the component bounds Eq. (3.44) describe for given one-point statistics a rectangle in the 21-dimensional space of elastic constants. The union of all of these rectangles delivers the first-order properties-closure of Adams et al. (2013). Analogously the zeroth-order bounds deliver the zeroth-order properties-closure. For given materials, the zeroth-order properties-closure encloses the first-order properties-closure. Naturally, this can be continued by taking into consideration high-order bounds to examine design possibilities in higher-order properties-closures. But it should be kept in mind that bounds deliver sharp statements about values which are *not* achievable. For given materials, the zeroth-order bounds give information about properties which can not be realized independently of the microstructure variation and reflect only a notion of what might be achievable by the materials. Taking into account the first-order properties-closure, material designers will have a better understanding of what is *not* achievable for the chosen materials for any variation of the one-point statistics and a better notion of what *might* be achievable by variation of higher-point-statistics of the microstructure.

3.4.4 HS variational principle, bounds and localization

HS variational principle. We now shortly recapitulate the Hashin-Shtrikman (HS) variational principle of Hashin and Shtrikman (1962) extended with eigenfields following Willis (1981).

We consider a material volume v with prescribed deformation on the boundary $\boldsymbol{u} = \bar{\varepsilon}\boldsymbol{x}\ \forall \boldsymbol{x} \in \partial v$. The stress potential W is given as in Eq. (3.29). The stress $\boldsymbol{\sigma} = \partial W/\partial \varepsilon = \mathbb{C}[\varepsilon] - \boldsymbol{s}$ can be reformulated with a constant $\mathbb{C}_0$ and a constant eigenfield $\boldsymbol{s}_0$ as $\boldsymbol{\sigma} = \mathbb{C}_0[\varepsilon] - \boldsymbol{s}_0 + \hat{\boldsymbol{\tau}}$ with $\hat{\boldsymbol{\tau}} = \hat{\mathbb{C}}[\varepsilon] - \hat{\boldsymbol{s}}, \hat{\mathbb{C}} = \mathbb{C} - \mathbb{C}_0, \hat{\boldsymbol{s}} = \boldsymbol{s} - \boldsymbol{s}_0$ and the solution of the problem given as in Eq. (3.12) and Eq. (3.13) but now in the RVE

$$\varepsilon = \bar{\varepsilon} - \mathbb{G}\{\hat{\boldsymbol{\tau}} - \boldsymbol{s}_0\} = \bar{\varepsilon} - \mathbb{G}\{\hat{\boldsymbol{\tau}}\}\ ,\ \mathbb{G}\{\hat{\boldsymbol{\tau}}\} = \int_v \mathbb{G}(\boldsymbol{x}, \boldsymbol{x}')[\hat{\boldsymbol{\tau}}(\boldsymbol{x}')]\mathrm{d}v'\ . \quad (3.75)$$

at what $\mathbb{G}$ maps constants to zero, see Willis (1981). The solution can be expressed now in terms of the exact stress polarization $\hat{\boldsymbol{\tau}}$ by eliminating $\varepsilon = \hat{\mathbb{C}}^{-1}[\hat{\boldsymbol{\tau}} + \hat{\boldsymbol{s}}]$ as

$$\boldsymbol{K}(\boldsymbol{\tau}) = \hat{\mathbb{C}}^{-1}[\boldsymbol{\tau}] + \mathbb{G}\{\boldsymbol{\tau}\} - \tilde{\varepsilon}\ , \quad \tilde{\varepsilon} = \bar{\varepsilon} - \hat{\mathbb{C}}^{-1}[\hat{\boldsymbol{s}}]\ , \qquad (3.76)$$

where $\boldsymbol{K}(\boldsymbol{\tau} = \hat{\boldsymbol{\tau}}) = \boldsymbol{0}$ (i.e., Eq. (3.75)) holds for the exact polarization $\hat{\boldsymbol{\tau}}$. For any other stress polarization $\boldsymbol{\tau}$, $\boldsymbol{K}(\boldsymbol{\tau}) \neq \boldsymbol{0}$, in general. Based on

$\boldsymbol{K}$, the Hashin-Shtrikman functional $F = F(\boldsymbol{\tau})$ including eigenfields is defined as follows

$$F = \left\langle \frac{1}{2}\boldsymbol{\tau} \cdot \hat{\mathbb{C}}^{-1}[\boldsymbol{\tau}] + \frac{1}{2}\boldsymbol{\tau} \cdot (\mathbb{G}\{\boldsymbol{\tau}\}) - \boldsymbol{\tau} \cdot \tilde{\boldsymbol{\varepsilon}} \right\rangle_v = \frac{1}{2}\langle \boldsymbol{K} \cdot \boldsymbol{\tau} \rangle_v - \frac{1}{2}\langle \boldsymbol{\tau} \cdot \tilde{\boldsymbol{\varepsilon}} \rangle_v \,, \tag{3.77}$$

which fulfills $\delta F = \langle \boldsymbol{K} \delta \boldsymbol{\tau} \rangle_v$, i.e., F is stationary for $\hat{\boldsymbol{\tau}}$, for which F has the value

$$\begin{aligned} \hat{F} &= F(\boldsymbol{\tau} = \hat{\boldsymbol{\tau}}) = \bar{W}_0 - \bar{W} \,, \\ \bar{W}_0 &= \frac{1}{2}\bar{\boldsymbol{\varepsilon}} \cdot \mathbb{C}_0[\bar{\boldsymbol{\varepsilon}}] - \bar{\boldsymbol{\varepsilon}} \cdot \boldsymbol{s}_0 - \frac{1}{2}(\langle k \rangle_v + \langle \hat{\boldsymbol{s}} \cdot \hat{\mathbb{C}}^{-1}[\hat{\boldsymbol{s}}] \rangle_v) \,. \end{aligned} \tag{3.78}$$

The second variation is given by

$$\delta^2 F = \langle \delta\boldsymbol{\tau} \cdot \hat{\mathbb{C}}^{-1}[\delta\boldsymbol{\tau}] + \delta\boldsymbol{\tau} \cdot (\mathbb{G}\{\delta\boldsymbol{\tau}\}) \rangle_v \,. \tag{3.79}$$

HS bounds. In Willis (1977) it has been shown, that, for positive definite $\mathbb{C}_0$ fulfilling $\mathbb{C} \leq \mathbb{C}_0$ (at every point of the RVE), the operator $\hat{\mathbb{C}}^{-1} + \mathbb{G}$ in the second variation of F is negative definite. This implies that the stationary value $\hat{F}$ is a maximum, i.e., $\hat{F} = \bar{W}_0 - \bar{W} \geq F(\boldsymbol{\tau}) \forall \boldsymbol{\tau}$. This property allows the generation of an upper bound W^{HS+} of $\bar{W}$ if the polarization field is constrained in some way, say, $\boldsymbol{\tau} \in C_\tau$ (where C_τ symbolizes a constraint set for $\boldsymbol{\tau}$). This delivers the upper bound

$$W^{HS+} = \bar{W}_0 - \max_{\boldsymbol{\tau} \in C_\tau} F(\boldsymbol{\tau}) \geq \bar{W} \,. \tag{3.80}$$

As shown in Willis (1977), analogous treatment can be followed for positive definite $\mathbb{C}_0$ fulfilling $\mathbb{C}_0 \leq \mathbb{C}$ in order to derive a lower bound

$$W^{HS-} = \bar{W}_0 - \min_{\boldsymbol{\tau} \in C_\tau} F(\boldsymbol{\tau}) \leq \bar{W} \,. \tag{3.81}$$

The HS bounds hold then for all effective strains $\bar{\varepsilon}$, i.e.,

$$W^{HS-} \leq \bar{W} \leq W^{HS+} \quad \forall \bar{\varepsilon} \, . \tag{3.82}$$

Based on Eq. (3.41), bounds for the complementary potential can be obtained with the Legendre-Fenchel transforms of the corresponding HS bounds

$$U^{HS-} = (W^{HS+})^* \leq \bar{U} \leq U^{HS+} = (W^{HS-})^* \, . \tag{3.83}$$

HS bounds for isotropic two-point statistics. For ellipsoidal two-point-statistics with no long-range order in a large volume and phase-wise constant stress polarization, to be denoted as $\boldsymbol{\tau} \in C_\tau$, the term $\langle \boldsymbol{\tau} \cdot (\mathbb{G}\{\boldsymbol{\tau}\}) \rangle_v$ is replaced with the operator $\mathbb{G}_\infty$ based on Green's function for the infinite body with vanishing fluctuation of the displacement field, see Willis (1977), by

$$\langle \boldsymbol{\tau} \cdot (\mathbb{G}\{\boldsymbol{\tau}\}) \rangle_v = \langle \boldsymbol{\tau} \cdot \mathbb{G}_\infty \{\boldsymbol{\tau} - \langle \boldsymbol{\tau} \rangle_v\} \rangle_v = \langle \boldsymbol{\tau} \cdot \mathbb{P}_0 [\boldsymbol{\tau} - \langle \boldsymbol{\tau} \rangle_v] \rangle_v \, . \tag{3.84}$$

The term $\boldsymbol{\tau} - \langle \boldsymbol{\tau} \rangle_v$ is used in order to achieve convergence of the integral concerning $\mathbb{G}_\infty$, see Willis (1977) and Willis (1981) for a detailed discussion. For ellipsoidal two-point statistics, the integral concerning $\mathbb{G}_\infty$ simplifies to an integral with the polarization tensor $\mathbb{P}_0$, see Willis (1977) and Willis (1981),

$$\mathbb{P}_0 = \int_{S_2} \tilde{\mathbb{P}}_0 \mathrm{d}n \, , \quad \tilde{\mathbb{P}}_0 = \frac{\mathbb{I}^S ((\mathbb{C}_0^{\mathsf{T_I}} [\boldsymbol{n}^{\otimes 2}])^{-1} \otimes (\boldsymbol{n}^{\otimes 2}))^{\mathsf{T_I}} \mathbb{I}^S}{4\pi \det(\boldsymbol{A}) \, \|\boldsymbol{A}^{-1} \boldsymbol{n}\|^3} \, , \tag{3.85}$$

where the symmetric $\boldsymbol{A}$ reflects the ellipsoidal shape of the two-point statistics (2-PPF). The eigenvalues of the symmetric $\boldsymbol{A}$ correspond to the inverse of the aspect ratios of the average inclusion of the corresponding phase. The symbol $\mathsf{T_I}$ represent the *inner* index transposition of a fourth-order tensor, defined in components of a fourth-order tensor $\mathbb{A}$ as

$A^{\mathsf{T}_\mathsf{I}}_{ijkl} = A_{ikjl}$. It should be remarked that the integrand $\tilde{\mathbb{P}}_0$ is minor and major symmetric even for a triclinic stiffness $\mathbb{C}_0$, such that $\mathbb{P}_0$ is minor and major symmetric. The position independent tensor $\mathbb{P}_0$ depends on the ellipsoidal shape of the two-point-statistics and on the reference stiffness $\mathbb{C}_0$ of the comparison medium. For isotropic two-point statistics, i.e., $\boldsymbol{A} = \boldsymbol{I}$, and anisotropic comparison stiffness $\mathbb{C}_0$, the analytic form of $\mathbb{P}_0$ is know only for some cases. If the two-point-statistics are chosen anisotropic but still ellipsoidal, then $\mathbb{P}_0$ has to be determined numerically for anisotropic $\mathbb{C}_0$. In this work, isotropic two-point statistics will be assumed and an isotropic comparison stiffness $\mathbb{C}_0 = c_1\mathbb{P}_1 + c_2\mathbb{P}_2$ will be used. For this choice, the tensor $\mathbb{P}_0$ is also isotropic and given by

$$\mathbb{P}_0 = \frac{1}{c_1 + 2c_2}\mathbb{P}_1 + \frac{2}{5c_2}\frac{c_1 + 3c_2}{c_1 + 2c_2}\mathbb{P}_2 \,. \tag{3.86}$$

This will be assumed from this point on for the rest of the present work. The HS functional Eq. (3.77) is reformulated as follows

$$F = \frac{1}{2}\langle \boldsymbol{\tau} \cdot \mathbb{D}[\boldsymbol{\tau}]\rangle_v - \frac{1}{2}\langle \boldsymbol{\tau}\rangle_v \cdot \mathbb{P}_0[\langle \boldsymbol{\tau}\rangle_v] - \langle \boldsymbol{\tau} \cdot \tilde{\varepsilon}\rangle_v \,, \quad \mathbb{D} = \hat{\mathbb{C}}^{-1} + \mathbb{P}_0 \,. \tag{3.87}$$

Stationarity of Eq. (3.87) with respect to the stress polarization demands

$$\mathbb{D}[\boldsymbol{\tau}] - \mathbb{P}_0[\langle \boldsymbol{\tau}\rangle_v] - \tilde{\varepsilon} = \mathbf{0} \,. \tag{3.88}$$

This might be solved first for $\langle \boldsymbol{\tau}\rangle_v$ and then for $\boldsymbol{\tau}$. The explicit computations are documented in Appendix D. The value of F for $\boldsymbol{\tau}$ fulfilling Eq. (3.88) is obtained as

$$F = -\frac{1}{2}\langle \boldsymbol{\tau} \cdot \tilde{\varepsilon}\rangle_v \,. \tag{3.89}$$

Since the phasewise constant stress polarization is linear in the strain $\tilde{\varepsilon} = \bar{\varepsilon} - \hat{\mathbb{C}}^{-1}[\hat{s}]$ (due to Eq. (3.88)), it is clear that Eq. (3.89) delivers a quadratic polynomial in $\bar{\varepsilon}$. The HS potential W^{HS} is obtained after some technical computations as

$$W^{HS} = \bar{W}_0 + \frac{1}{2}\langle \boldsymbol{\tau} \cdot \tilde{\varepsilon} \rangle_v = \frac{1}{2}\bar{\varepsilon} \cdot \mathbb{C}^{HS}[\bar{\varepsilon}] - \bar{\varepsilon} \cdot \boldsymbol{s}^{HS} - \frac{1}{2}k^{HS} , \tag{3.90}$$

which based on the minor and major symmetric tensor

$$\mathbb{L} = [\mathbb{C} - (\mathbb{C}_0 - \mathbb{P}_0^{-1})]^{-1} \tag{3.91}$$

possesses the material properties, see Appendix D

$$\begin{aligned} \mathbb{C}^{HS} &= \mathbb{C}_0 - \mathbb{P}_0^{-1} + \langle \mathbb{L} \rangle_v^{-1} , \quad \boldsymbol{s}^{HS} = \langle \mathbb{L} \rangle_v^{-1} [\langle \mathbb{L}[\boldsymbol{s}] \rangle_v] , \\ k^{HS} &= \langle k \rangle_v + \langle \boldsymbol{s} \cdot \mathbb{L}[\boldsymbol{s}] \rangle_v - \boldsymbol{s}^{HS} \cdot \langle \mathbb{L}[\boldsymbol{s}] \rangle_v . \end{aligned} \tag{3.92}$$

It is remarked that $\boldsymbol{s}_0$ vanishes completely in these results, i.e., $\boldsymbol{s}_0$ has no influence on the HS bounds for isotropic two-point statistics if the polarization field is determined by the stationarity condition. An upper HS bound is obtained as $W^{HS} = W^{HS+}$ using a comparison stiffness fulfilling $\mathbb{C} \leq \mathbb{C}_0$ (and $W^{HS} = W^{HS-}$ for $\mathbb{C} \geq \mathbb{C}_0$).

Comments on optimal HS bounds. The condition $\mathbb{C} \leq \mathbb{C}_0$ for an upper HS bound describes for given $\mathbb{C}$ a set $\mathcal{C}_0^+$ for the components or eigenvalues of $\mathbb{C}_0$. Every point in $\mathcal{C}_0^+$ generates an upper HS bound W^{HS+} for which $\bar{W} \leq W^{HS+} \forall \bar{\varepsilon}$ holds. Without further specifications, we are not able to choose a point from $\mathcal{C}_0^+$ which would deliver the smallest upper HS bound. The smallest upper HS bound is obtained by minimizing the HS expression with respect to $\mathcal{C}_0^+$, i.e., the optimal HS upper bound $\hat{W}^{HS+}$ is defined as

$$\hat{W}^{HS+} = \min_{\mathbb{C}_0 \in \mathcal{C}_0^+} W^{HS} . \tag{3.93}$$

This optimization problem is highly nonlinear in $\mathbb{C}_0$ and analytic solutions are not known, in general. The minimizing $\mathbb{C}_0$ depends, in general, on all material properties of the phases, on their volume fractions and on $\bar{\varepsilon}$. The corresponding HS material properties depend naturally also on these quantities such that, e.g., the corresponding $\hat{\mathbb{C}}^{HS}$ would have to be analyzed for all varying quantities.

A much simpler, but less optimal, upper HS bound can be obtained by using the upper zeroth-order bound of the stiffness, which fulfills the condition $\mathbb{C} \leq \mathbb{C}_0$, i.e.,

$$W^{2+} = W^{HS}\Big|_{\mathbb{C}_0 = \mathbb{C}^{0+}} , \tag{3.94}$$

with corresponding properties $\mathbb{C}^{2+}$, s^{2+} and k^{2+}. This upper bound is less optimal, i.e., $\bar{W} \leq \hat{W}^{HS+} \leq W^{2+} \forall \bar{\varepsilon}$, but it offers the possibility to analyze all corresponding HS material properties and derive bounds in closed forms for all effective material properties.

It is shortly remarked that analogous arguments can be followed for the lower optimal HS bound and the lower HS bound defined with $\mathbb{C}^{0-}$, denoted as $W^{2-} = W^{HS}|_{\mathbb{C}_0 = \mathbb{C}^{0-}}$ with corresponding $\mathbb{C}^{2-}$, s^{2-} and k^{2-}.

Localization tensors for HS. The HS stiffness might be reformulated as follows

$$\begin{aligned} \mathbb{C}^{HS} &= [(\mathbb{C}_0 - \mathbb{P}_0^{-1})\langle \mathbb{L} \rangle_v + \mathbb{I}^S]\langle \mathbb{L} \rangle_v^{-1} \\ &= \langle [(\mathbb{C}_0 - \mathbb{P}_0^{-1}) + \mathbb{L}^{-1}]\mathbb{L} \rangle_v \langle \mathbb{L} \rangle_v^{-1} \\ &= \langle \mathbb{C}\mathbb{A}^{HS} \rangle_v \end{aligned} \tag{3.95}$$

with

$$\mathbb{A}^{HS} = \mathbb{L}\langle \mathbb{L} \rangle_v^{-1} . \tag{3.96}$$

The gradient of the HS potential W^{HS} with respect to $\bar{\varepsilon}$ offers an estimate for the effective stress, to be denoted as $\bar{\boldsymbol{\sigma}}^{HS}$, which can be expressed as

$$\begin{aligned}\bar{\boldsymbol{\sigma}}^{HS} &= \frac{\partial W^{HS}}{\partial \bar{\boldsymbol{\varepsilon}}} = \mathbb{C}^{HS}[\bar{\boldsymbol{\varepsilon}}] - \boldsymbol{s}^{HS} \\ &= \langle \mathbb{C}\mathbb{A}^{HS}[\bar{\boldsymbol{\varepsilon}}] - (\mathbb{A}^{HS})^{\mathsf{T}}[\boldsymbol{s}]\rangle_v \\ &= \langle \mathbb{C}[\mathbb{A}^{HS}[\bar{\boldsymbol{\varepsilon}}] + \boldsymbol{a}^{HS}] - \boldsymbol{s}\rangle_v\end{aligned} \tag{3.97}$$

with

$$\boldsymbol{a}^{HS} = \mathbb{C}^{-1}(\mathbb{I}^S - (\mathbb{A}^{HS})^{\mathsf{T}})[\boldsymbol{s}] \,. \tag{3.98}$$

These reformulations motivate the HS localization relations for the local fields $\boldsymbol{\varepsilon}^{HS}$ and $\boldsymbol{\sigma}^{HS}$

$$\begin{aligned}\boldsymbol{\varepsilon}^{HS} &= \mathbb{A}^{HS}[\bar{\boldsymbol{\varepsilon}}] + \boldsymbol{a}^{HS} \,, \\ \boldsymbol{\sigma}^{HS} &= \mathbb{C}[\boldsymbol{\varepsilon}^{HS}] - \boldsymbol{s} = \mathbb{C}\mathbb{A}^{HS}[\bar{\boldsymbol{\varepsilon}}] - (\mathbb{A}^{HS})^{\mathsf{T}}[\boldsymbol{s}] \,.\end{aligned} \tag{3.99}$$

These localization relations allow to estimate and investigate the local fields in the composite based on the HS results. These relations are summarized as

$$\begin{aligned}&\mathbb{A}^{HS} = \mathbb{L}\langle\mathbb{L}\rangle_v^{-1} \,, \\ &\boldsymbol{a}^{HS} = \mathbb{C}^{-1}(\mathbb{I}^S - (\mathbb{A}^{HS})^{\mathsf{T}})[\boldsymbol{s}] \,, \\ &\boldsymbol{\varepsilon}^{HS} = \mathbb{A}^{HS}[\bar{\boldsymbol{\varepsilon}}] + \boldsymbol{a}^{HS} \,, \\ &\boldsymbol{\sigma}^{HS} = \mathbb{C}[\boldsymbol{\varepsilon}^{HS}] - \boldsymbol{s} = \mathbb{C}\mathbb{A}^{HS}[\bar{\boldsymbol{\varepsilon}}] - (\mathbb{A}^{HS})^{\mathsf{T}}[\boldsymbol{s}] \,, \\ &\bar{\boldsymbol{\sigma}}^{HS} = \frac{\partial W^{HS}}{\partial \bar{\boldsymbol{\varepsilon}}} = \mathbb{C}^{HS}[\bar{\boldsymbol{\varepsilon}}] - \boldsymbol{s}^{HS} = \langle \boldsymbol{\sigma}^{HS}\rangle_v \,.\end{aligned} \tag{3.100}$$

It is remarked that the implied localization relation is *not* normalized as expected since $\langle \mathbb{A}^{HS}\rangle_v = \mathbb{I}^S$ holds but $\langle \boldsymbol{a}^{HS}\rangle_v \neq \mathbf{0}$. This may be unexpected but it a property implied by the polarization field determined by the stationarity condition. Naturally, one may use other approaches to formulate a sensible polarization field, see, e.g., Willis (1981), but it remains then unclear how far away such approaches are from the stationary value of the HS functional, which may deliver less optimal bounds.

Dual principle, bounds and localization. The equation $\varepsilon = \bar{\varepsilon} - \mathbb{G}\{\hat{\tau}\}$ can be reformulated based on $\sigma = \mathbb{C}_0[\varepsilon] + \hat{\tau}, \mathbb{S}_0 = \mathbb{C}_0^{-1}$ and $\varepsilon = \mathbb{S}_0[\sigma] + \hat{\xi}$ in terms of the exact strain polarization $\hat{\xi} = -\mathbb{S}_0[\hat{\tau}]$ as

$$\sigma = \bar{\sigma} - \mathbb{T}\{\hat{\xi}\}\,, \quad \mathbb{T}\{\hat{\xi}\} = \mathbb{C}_0[\hat{\xi} - \langle\hat{\xi}\rangle_v - \mathbb{G}\{\mathbb{C}_0[\hat{\xi}]\}]\,, \tag{3.101}$$

where $-\mathbb{T}\{\hat{\xi}\}$ is a statically admissible stress with zero mean. Now, for a heterogeneous material with $\varepsilon = \mathbb{S}[\sigma] + e$ the exact strain polarization is given by $\hat{\xi} = \hat{\mathbb{S}}[\sigma] + e$ with $\hat{\mathbb{S}} = \mathbb{S} - \mathbb{S}_0$. Following the HS variational principle construction, the relation Eq. (3.101) can be described in terms of an arbitrary strain polarization ξ

$$H(\xi) = \hat{\mathbb{S}}^{-1}[\xi] + \mathbb{T}\{\xi\} - \tilde{\sigma}\,, \quad \tilde{\sigma} = \bar{\sigma} + \hat{\mathbb{S}}[e]\,, \quad H(\hat{\xi}) = \mathbf{0}\,. \tag{3.102}$$

The corresponding functional

$$J = \frac{1}{2}\langle\xi \cdot \hat{\mathbb{S}}^{-1}[\xi]\rangle_v + \frac{1}{2}\langle\xi \cdot \mathbb{T}\{\xi\}\rangle_v - \langle\xi \cdot \tilde{\sigma}\rangle_v \tag{3.103}$$

is stationary at $\xi = \hat{\xi}$ and has the value

$$\hat{J} = J(\hat{\xi}) = \bar{U}_0 - \bar{U}\,, \quad \bar{U}_0 = \frac{1}{2}\bar{\sigma} \cdot \mathbb{S}_0[\bar{\sigma}] + \frac{1}{2}(\langle l\rangle_v - \langle e \cdot \hat{\mathbb{S}}[e]\rangle_v)\,. \tag{3.104}$$

Based on the second variation of J and following analogous reasoning as proved in Willis (1977), it can be shown that for $\mathbb{S} \le \mathbb{S}_0$ the operator $\hat{\mathbb{S}}^{-1} + \mathbb{T}$ is negative definite, such that the stationary value is a maximum and a corresponding HS upper bound U^{HS+} can be obtained. Analogous reasoning, following Willis (1977), can be applied for $\mathbb{S}_0 \le \mathbb{S}$ in order to generate the lower bound U^{HS-}.

Now, the classical HS bounds for ellipsoidal 2-point statistics with $\mathbb{P}_0$ are obtained by replacing

$$
\begin{aligned}
&\langle \boldsymbol{\xi} \cdot \mathbb{T}\{\boldsymbol{\xi}\}\rangle_v \\
&= \langle \boldsymbol{\xi} \cdot \mathbb{C}_0[\boldsymbol{\xi}]\rangle_v - \langle \boldsymbol{\xi}\rangle_v \cdot \mathbb{C}_0[\langle \boldsymbol{\xi}\rangle_v] - \langle (\mathbb{C}_0[\boldsymbol{\xi}]) \cdot \mathbb{G}\{\mathbb{C}_0[\boldsymbol{\xi}]\}\rangle_v \\
&= \langle \boldsymbol{\xi} \cdot \mathbb{C}_0[\boldsymbol{\xi}]\rangle_v - \langle \boldsymbol{\xi}\rangle_v \cdot \mathbb{C}_0[\langle \boldsymbol{\xi}\rangle_v] - \langle (\mathbb{C}_0[\boldsymbol{\xi}]) \cdot \mathbb{P}_0[\mathbb{C}_0[\boldsymbol{\xi}] - \langle \mathbb{C}_0[\boldsymbol{\xi}]\rangle_v]\rangle_v \\
&= \langle \boldsymbol{\xi} \cdot \mathbb{Q}_0[\boldsymbol{\xi}]\rangle_v - \langle \boldsymbol{\xi}\rangle_v \cdot \mathbb{Q}_0[\langle \boldsymbol{\xi}\rangle_v]
\end{aligned}
\tag{3.105}
$$

with the constant minor and major symmetric tensor

$$
\mathbb{Q}_0 = \mathbb{C}_0 - \mathbb{C}_0\mathbb{P}_0\mathbb{C}_0 \, . \tag{3.106}
$$

The functional J is expressed now as

$$
J = \frac{1}{2}\langle \boldsymbol{\xi} \cdot (\hat{\mathbb{S}}^{-1} + \mathbb{Q}_0)[\boldsymbol{\xi}]\rangle_v - \frac{1}{2}\langle \boldsymbol{\xi}\rangle_v \cdot \mathbb{Q}_0[\langle \boldsymbol{\xi}\rangle_v] - \langle \boldsymbol{\xi} \cdot \tilde{\boldsymbol{\sigma}}\rangle_v \, . \tag{3.107}
$$

The algebraic structure is naturally identical to Eq. (3.87), such that we can immediately present the dual HS potential based on the minor and major symmetric tensor $\mathbb{M}$

$$
\begin{aligned}
U^{HS} &= \tfrac{1}{2}\bar{\boldsymbol{\sigma}} \cdot \mathbb{S}^{HS}[\bar{\boldsymbol{\sigma}}] + \bar{\boldsymbol{\sigma}} \cdot \boldsymbol{e}^{HS} + \tfrac{1}{2}l^{HS} \, , \quad \mathbb{M} = [\mathbb{S} - (\mathbb{S}_0 - \mathbb{Q}_0^{-1})]^{-1} \, , \\
\mathbb{S}^{HS} &= \mathbb{S}_0 - \mathbb{Q}_0^{-1} + \langle \mathbb{M}\rangle_v^{-1} \, , \quad \boldsymbol{e}^{HS} = \langle \mathbb{M}\rangle_v^{-1}[\langle \mathbb{M}[\boldsymbol{e}]\rangle_v] \, , \\
l^{HS} &= \langle l\rangle_v - \langle \boldsymbol{e} \cdot \mathbb{M}[\boldsymbol{e}]\rangle_v + \boldsymbol{e}^{HS} \cdot \langle \mathbb{M}[\boldsymbol{e}]\rangle_v \, .
\end{aligned}
\tag{3.108}
$$

Further computation shows that $U^{HS} = (W^{HS})^*$ holds. The advantage of Eq. (3.108) is that the corresponding bounds are given explicitly in terms of $\mathbb{S}, \boldsymbol{e}$ and l. This might be useful for problems starting with complementary energy formulations. One example for viscoplastic polycrystals will be discussed in Chapter 4.

It should be remarked that analogous arguments for optimal and simpler upper and lower bounds, following Eq. (3.93) and Eq. (3.94), can be considered.

Analogous localization relations, as the ones given in Eq. (3.99), are obtained for the dual formulation, expressed as

$$\begin{aligned} \mathbb{B}^{HS} &= \mathbb{M}\langle\mathbb{M}\rangle_v^{-1} , \\ \boldsymbol{b}^{HS} &= \mathbb{S}^{-1}((\mathbb{B}^{HS})^\mathsf{T} - \mathbb{I}^S)[\boldsymbol{e}] , \\ \boldsymbol{\sigma}^{HS} &= \mathbb{B}^{HS}[\bar{\boldsymbol{\sigma}}] + \boldsymbol{b}^{HS} , \\ \boldsymbol{\varepsilon}^{HS} &= \mathbb{S}[\boldsymbol{\sigma}^{HS}] + \boldsymbol{e} = \mathbb{S}\mathbb{B}^{HS}[\bar{\boldsymbol{\sigma}}] + (\mathbb{B}^{HS})^\mathsf{T}[\boldsymbol{e}] , \\ \bar{\boldsymbol{\varepsilon}}^{HS} &= \frac{\partial U^{HS}}{\partial \bar{\boldsymbol{\sigma}}} = \mathbb{S}^{HS}[\bar{\boldsymbol{\sigma}}] + \boldsymbol{e}^{HS} = \langle \boldsymbol{\varepsilon}^{HS} \rangle_v . \end{aligned} \tag{3.109}$$

Classical limits and first-order bounds. For the comparison stiffness $\mathbb{C}_0 = \gamma\hat{\mathbb{C}}_0, \mathbb{P}_0(\mathbb{C}_0) = \frac{1}{\gamma}\mathbb{P}_0(\hat{\mathbb{C}}_0) = \frac{1}{\gamma}\hat{\mathbb{P}}_0$ and $\mathbb{P}_0^{-1} = \gamma\hat{\mathbb{P}}_0^{-1}$ follow. The limit for vanishing γ of W^{HS} delivers with $\mathbb{C}^{-1} = \mathbb{S}, \boldsymbol{e} = \mathbb{C}^{-1}[\boldsymbol{s}]$ and $l = k + \boldsymbol{s} \cdot \mathbb{C}^{-1}[\boldsymbol{s}]$

$$\lim_{\gamma\to 0} W^{HS} = \frac{1}{2}\bar{\boldsymbol{\varepsilon}} \cdot \langle\mathbb{S}\rangle_v^{-1}[\bar{\boldsymbol{\varepsilon}}] - \bar{\boldsymbol{\varepsilon}} \cdot \langle\mathbb{S}\rangle_v^{-1}[\langle\boldsymbol{e}\rangle_v] - \frac{1}{2}(\langle l\rangle_v - \langle\boldsymbol{e}\rangle_v \cdot \langle\mathbb{S}\rangle_v^{-1}[\langle\boldsymbol{e}\rangle_v]) . \tag{3.110}$$

The Legendre-Fenchel transform of this limit equals to

$$\begin{aligned} (\lim_{\gamma\to 0} W^{HS})^* &= \sup_{\bar{\boldsymbol{\varepsilon}}}\{\bar{\boldsymbol{\varepsilon}} \cdot \bar{\boldsymbol{\sigma}} - \lim_{\gamma\to 0} W^{HS}\} \\ &= \tfrac{1}{2}\bar{\boldsymbol{\sigma}} \cdot \langle\mathbb{S}\rangle_v[\bar{\boldsymbol{\sigma}}] + \bar{\boldsymbol{\sigma}} \cdot \langle\boldsymbol{e}\rangle_v + \tfrac{1}{2}\langle l\rangle_v \end{aligned} \tag{3.111}$$

such that this limit is in accordance with the upper first-order bounds in the complementary space.

For the complementary HS potential, the limit for $\gamma \to \infty$ is computed based on $\mathbb{S}_0 = \mathbb{C}_0^{-1} = \frac{1}{\gamma}\hat{\mathbb{C}}_0$ as

$$\begin{aligned} \lim_{\gamma\to\infty} U^{HS} &= \frac{1}{2}\bar{\boldsymbol{\sigma}} \cdot \langle \mathbb{C} \rangle_v^{-1}[\bar{\boldsymbol{\sigma}}] + \bar{\boldsymbol{\sigma}} \cdot \langle \mathbb{C} \rangle_v^{-1}[\langle \boldsymbol{s} \rangle_v] \\ &\quad + \frac{1}{2}(\langle k \rangle_v + \langle \boldsymbol{s} \rangle_v \cdot \langle \mathbb{C} \rangle_v^{-1}[\langle \boldsymbol{s} \rangle_v]) . \end{aligned} \tag{3.112}$$

Its Legendre-Fenchel transform yields

$$\begin{aligned} (\lim_{\gamma\to\infty} U^{HS})^* &= \sup_{\bar{\boldsymbol{\sigma}}}\{\bar{\boldsymbol{\varepsilon}} \cdot \bar{\boldsymbol{\sigma}} - \lim_{\gamma\to\infty} U^{HS}\} \\ &= \tfrac{1}{2}\bar{\boldsymbol{\varepsilon}} \cdot \langle \mathbb{C} \rangle_v[\bar{\boldsymbol{\varepsilon}}] - \bar{\boldsymbol{\varepsilon}} \cdot \langle \boldsymbol{s} \rangle_v - \tfrac{1}{2}\langle k \rangle_v \end{aligned} \tag{3.113}$$

such that this limit is in accordance with the upper first-order bound of $\bar{W}$. It is remarked that these limits are implied by the polarization field determined through the stationarity condition (3.88). It is unclear if other polarization fields (not determined by the stationarity condition) deliver expressions which also deliver these limits.

Comments on HS expressions and texture dependency. The expressions for the HS material properties given Eq. (3.92) and Eq. (3.108) offer the most clear and computationally advantageous form possible for the general anisotropic case. Compared to the algebraically identical expression for the HS stiffness given in Willis (1977), the HS expression for the stiffness given in Eq. (3.92), also obtained already by Walpole (1966) for pure linear elasticity, shows immediately that the HS stiffness is symmetric for arbitrary anisotropic constituents.

Further, the HS stiffness of Eq. (3.92) requires only one volume average of major symmetric tensors over the RVE, compared to 2 averages of in general not major symmetric tensors in the expression of Willis (1977).

Due to the major symmetry of $\mathbb{L}$ (and $\mathbb{M}$), it is immediately visible that for all elastic-anisotropy classes, the average $\langle \mathbb{L} \rangle_v$ (and $\langle \mathbb{M} \rangle_v$) and consequently all HS material properties depend in terms of the texture *only on second- and fourth-order texture coefficients of the corresponding CODFs of each phase.* These are the very same relevant microstructural texture variables of the first-order bounds. This result seems not to be known in literature and offers a new insight into the present work. This result is also obtainable through the expression of Willis (1977) but only by considering the harmonic decomposition of minor but not major symmetric tensors and the results of Lobos et al. (2017). There, it has been shown, that, for the special, not major symmetric tensors appearing in the HS stiffness expression of Willis (1977), the harmonic decomposition delivers vanishing first- and third-order harmonic parts. This result of Lobos et al. (2017) also implies that the corresponding orientation averages depend only on second- and fourth-order texture coefficients, but in a much more complicated fashion which is algebraically equivalent to the much simpler results of this work.

Finally, it should be remarked that, as discussed in Section 1.1, for a polycrystal each crystallographic orientation is considered as a distinct phase. This means that, in the case of polycrystalline materials, the phasewise constant polarization, isotropic two-point-correlation and lack of long-range-order in the HS bounds refers to each crystallographic orientation of each material constituent.

3.4.5 Approximations based on HS expressions

Isotropic self-consistent approximation. By solving the elastic problem in a comparison medium with homogeneous stiffness $\mathbb{C}_0$ one might consider the comparison medium as the medium exactly with the effective stiffness $\mathbb{C}_0 = \bar{\mathbb{C}}$. The exact solution of the elastic problem allows this, since the exact solution is valid for arbitrary $\mathbb{C}_0$. This delivers an implicit equation for the effective stiffness. Any approximation of the comparison-medium-dependent localization operator $\mathbb{A}$, depending on the assumptions and motivation of the respective model, delivers a different implicit equation for $\bar{\mathbb{C}}$. These approximations are referred to as self-consistent approximations, see Hershey (1954), Kröner (1958) and Hill (1965). This approach has been extended in several linear and nonlinear applications, see, e.g., Willis (1977), Talbot and Willis (1987), Lebensohn et al. (2004), Knezevic et al. (2013). However, these approaches are costly in their computation since the implicit nonlinear equations require in general the computation of the tensor $\mathbb{G}$ for general anisotropic comparison stiffness $\mathbb{C}_0 = \bar{\mathbb{C}}$ and general microstructure statistics.

In this work, several simplifications are made. For the statistical assumptions considered for the HS bounds, a semi self-consistent approach is build upon the HS stiffness $\mathbb{C}^{HS}$. Even for a multiphase material, the crystallographic isotropic case, i.e., $f_p(\boldsymbol{Q}) = 1 \forall p$, is a manageable expression, since only the isotropic part of the "single crystal" tensor $\mathbb{L}$ (or $\mathbb{M}$ for the dual formulations) needs to be extracted and superimposed with the phase concentrations c_p with $p = 1, \ldots, m$. The isotropic self-consistent solution $\mathbb{C}_0^{SC} = c_1^{SC}\mathbb{P}_1 + c_2^{SC}\mathbb{P}_2$ of the self-consistent equation based on $\mathbb{C}^{HS}$ is noted in this work as

$$\mathbb{C}_0^{SC} = \mathbb{C}^{HS}|_{\mathbb{C}_0 = \mathbb{C}_0^{SC}, f_p = 1}, \tag{3.114}$$

and is determined by only two scalar nonlinear equations for the respective eigenvalues of isotropic stiffness tensors. It should be noted, that these equations still depend on the phase concentrations c_p. For single-phase polycrystalline materials, i.e., $m = 1$ and $c_1 = 1$ hold, Eq. (3.114) can be solved immediately by numeric computation. For multiphase materials, Eq. (3.114) can only be used implicitly. The isotropic stiffness $\mathbb{C}_0^{SC}$ is then used in the present work as a comparison material for W^{HS} in order to generate the isotropically self-consistent (ISC) properties from Eq. (3.92)

$$\{\mathbb{C}^{ISC}, \boldsymbol{s}^{ISC}, k^{ISC}, \mathbb{A}^{ISC}, \boldsymbol{a}^{ISC}\} = \{\mathbb{C}^{HS}, \boldsymbol{s}^{HS}, k^{HS}, \mathbb{A}^{HS}, \boldsymbol{a}^{HS}\}|_{\mathbb{C}_0=\mathbb{C}_0^{SC}} \tag{3.115}$$

for general crystallographic texture. This approach is not as accurate as the full self-consistent approach but offers a fast and explicit computation regarding the texture since, for isotropic $\mathbb{C}_0$, the tensor $\mathbb{P}_0$ is given analytically and the representations of averages for general texture in terms of texture coefficients, see Chapter 2, are directly applicable. Therefore, the isotropically self-consistent approximation Eq. (3.115) offers a practical approximation which is also usable for inverse problems as in materials design, where the full self-consistent approach would be highly cumbersome.

Naturally, dual properties of the complementary space can also be used.

Singular approximation. The obvious practical approximation based on the HS expression is to choose a "sensible" comparison material $\mathbb{C}_0$, e.g., located somewhere between the reference tensors delivering the corresponding upper and lower bounds. This approach is purely pragmatic and is, after long but simple computations, algebraically equivalent to the singular approximation of Fokin (1972), see also, e.g., Morawiec (1996) and Morawiec (2004) for the purely elastic case.

The simplest and a probably "sensible" comparison material for a multiphase material is the Hill average of the isotropic first-order bounds of the linear elastic properties. This approach noted in this work as SA1

$$\begin{aligned}&\{\mathbb{C}^{SA1}, \boldsymbol{s}^{SA1}, k^{SA1}, \mathbb{A}^{SA1}, \boldsymbol{a}^{SA1}\}\\&= \{\mathbb{C}^{HS}, \boldsymbol{s}^{HS}, k^{HS}, \mathbb{A}^{HS}, \boldsymbol{a}^{HS}\}|_{\mathbb{C}_0=\frac{1}{2}(\mathbb{C}^{1+}+\mathbb{C}^{1-})|_{f_p=1}}\,.\end{aligned} \tag{3.116}$$

This pragmatic approach offers an explicit computation in terms of the texture and phase concentrations even for multiphase materials and might be considered for cumbersome cases, where implementation of $\mathbb{C}^{ISC}$ require too many resources.

Again, dual properties of the complementary space can also be used.

3.5 Applications for linear thermomechanical properties

Assumptions and exact relations. For classical linear thermomechanical behavior, it is assumed that the deformations are small, that the mass density ρ is constant, and that the absolute temperature θ differs only slightly of a reference temperature θ_0, i.e., $\theta \approx \theta_0$. The eigenfield $\boldsymbol{s}$ in Section 3.3 corresponds to the thermal stresses $\boldsymbol{s} = \boldsymbol{\beta}\Delta\theta$ with the temperature deviation $\Delta\theta = \theta - \theta_0$ and the thermal stress coefficient tensor $\boldsymbol{\beta}$. The corresponding scalar k is expressed as $k = f\Delta\theta^2$, where f is connected to the classical heat capacity for constant strain c_ε, i.e., $f = \rho c_\varepsilon/\theta_0$. In the present work θ and θ_0 are assumed as homogeneous over the RVE. The effective free energy of the RVE is expressed following Eq. (3.29) as

$$\bar{W} = \frac{1}{2}\bar{\varepsilon} \cdot \bar{\mathbb{C}}[\bar{\varepsilon}] - \bar{\varepsilon} \cdot \bar{\boldsymbol{\beta}}\Delta\theta - \frac{1}{2}\bar{f}\Delta\theta^2 \,. \tag{3.117}$$

Bounds, properties-closure and approximations. It is clear that the thermoelastic properties can be bounded with the zeroth-, first- and the respective second-order HS bounds (for assumed statistics). For MPPMs, the texture enters the first- and second-order bounds through the second- and fourth-order texture coefficients in all expressions.

The component bounds of the stiffness *and* of the effective thermal stress (and expansion) coefficient can be investigated with the aid of the zeroth-, first- and second-order properties-closure for varying texture. These maps will offer valuable information to materials designer in several applications.

Approximations for SPPMs can be built with the ISC approach. For MPPMs, the SA1 approach might be an option for pragmatic homogenization applications.

3.6 Applications for physically nonlinear problems

Variational principles based on linear comparison composite (LCC). For physically nonlinear problems, bounds and estimates can be generated, as introduced by Ponte Castañeda (1991), by taking into consideration a comparison linear composite. Based on the work of Ponte Castañeda (1991), several extensions for nonlinear materials have been developed by the use of the LCC for the generation of bounds and self-consistent estimates, see, e.g., DeBotton and Ponte Castañeda (1995), Ponte Castañeda (1996), Ponte Castañeda and Suquet (1997), Liu and Ponte Castañeda (2004), Ponte Castañeda et al. (2004), Ponte Castañeda (2015) and Ponte Castañeda (2016).

In this section, we consider the core idea of Ponte Castañeda (1991) in an RVE with local elastic energy density W with growth larger than quadratic in terms of the strains. This will be noted as $W(\varepsilon) > O(\varepsilon^2)$. As introduced in Ponte Castañeda (1991), we use a linear comparison composite (LCC) with local elastic energy density W_c of quadratic growth, $W_c(\varepsilon) = O(\varepsilon^2)$. Based solely on these properties the following auxiliary quantity is obtained

$$V_c = \sup_{\boldsymbol{\varepsilon}}\{W_c - W\} . \tag{3.118}$$

The definition of V_c implies the inequality

$$V_c \geq W_c - W \quad \forall \boldsymbol{\varepsilon} \tag{3.119}$$

at every point $\boldsymbol{x}$ of the RVE. The effective energy density $\bar{W}$ is defined through the solution of the boundary value problem which equals the infimum with respect to the displacement being kinematically compatible, i.e., $\bar{W} = \inf_{\boldsymbol{u}\in K}\langle W\rangle_v$. Rearrangement of Eq. (3.119), integration over the RVE and consideration of the infimum with respect to K on both sides yields

$$\bar{W} \geq \bar{W}_c - \bar{V}_c \,, \quad \bar{W}_c = \inf_{\boldsymbol{u}\in K} \langle W_c\rangle_v \,, \quad \bar{V}_c = \langle V_c\rangle_v \tag{3.120}$$

The effective energy density $\bar{W}_c$ is the exact effective energy density of the LCC, a physically linear material. For example, the comparison potential W_c could have been formulated as $W_c = \frac{1}{2}\boldsymbol{\varepsilon}\cdot\mathbb{C}_c[\boldsymbol{\varepsilon}] - \boldsymbol{\varepsilon}\cdot\boldsymbol{s}_c - \frac{1}{2}k_c$. It should be shortly remarked, that this method is analogously applicable to other physical problems, as to be illustrated in Section 4.3.2 for viscoplastic behavior.

Application of physically linear expressions. For $\bar{W}_c$ all previous expressions for linear elastic (and linear problems with eigenfields) of this work can be considered. For example, if a purely elastic comparison linear composite is taken into account, then, either with the Reuss bound or the lower HS bound, lower bounds of $\bar{W}_c$ can be given explicitly for multiphase polycrystalline materials. If an estimate is needed in practice, then, for multiphase polycrystalline materials, the isotropically self-consistent estimate might deliver a viable option. Improvements can be achieved by the usage of estimates of HS type with eigenfields, see, e.g., Ponte Castañeda (1996), Ponte Castañeda and Suquet (1997), Ponte Castañeda (2002) and Ponte Castañeda et al. (2004). The texture dependent HS expressions of Section 3.4.4 extend the applications of formulations with eigenfields if the eigenfields ($\boldsymbol{s}$ and $\boldsymbol{e}$) are assumed to have the same orientation at every material point as the local stiffness.

3.7 Review

In this chapter, the basic probabilistic description of multiphase random materials has been illustrated along with the basic concept of homogenization of mechanical problems with eigenfields. For linear problems with eigenfields, zeroth-, first- and second-order bounds of the effective material properties have been discussed. The second-order HS bounds have been illustrated depending on phase concentrations and texture coefficients. The isotropically self-consistent approximation has been proposed as a pragmatic explicit and efficient approximation in terms of finite-dimensional microstructural variables (phase concentrations and texture coefficients). This approximation and the SA1 approach are assumed to be useful for pragmatic applications in materials design. The application of the obtained expressions for thermomechanical materials and physically nonlinear bounds has been sketched. Explicit examples will be presented in the upcoming chapter.

Chapter 4

Applications in materials design and homogenization of textured materials

4.1 Overview

In this chapter, examples for the materials design of textured materials based on the expressions of the present work are illustrated. The chapter begins with a discussion of the main procedure and the terminology. Subsequently, orthotropic polycrystals of cubic single crystals are taken into consideration, and the elastic properties are designed in the first example. In the second example, the linear thermoelastic properties of transversely isotropic particle reinforced materials are investigated. The last example shows how viscoplastic properties of a textured polycrystal are bounded with a linear material and the corresponding HS expressions of the present work.

4.2 Materials design of linear properties

4.2.1 Procedure

Materials design problems are inverse problems. One of today's materials challenges is not to investigate several realizations of different compounds and tabulate measured data, but to construct the microstructure of a material, precisely based on demanded properties p^* with given tolerances T^* (properties-profile). Therefore, models incorporating the impact of microstructure m on interesting properties p up to a certain degree of accuracy are needed. First, depending on the intrinsic material anisotropy strength, some microstructure changes might or might not significantly influence the properties of interest. For example, the elastic properties of single crystal aluminum are cubic but almost isotropic, such that aluminum polycrystals show very similar elastic properties for all microstructures. Second, for materials showing strong anisotropy, depending on the strength of microstructure influence, it might be sufficient to incorporate only certain statistical moments of the microstructure, denoted schematically as $m_1, m_2, \ldots$ belonging to a microstructure variable set M.

The materials design is carried out in the present work as follows:

1. *Materials screening.*

   For bounded properties, we carry out a materials selection based on material databases, required properties p^* and tolerances T^*.

2. *Microstructure optimization.*

   For the chosen material, the first- and second-order properties-closures for the properties of interest are explored, based on phase volume fractions and texture variables. Then, a "sensible" approximation for the properties of interest, schematically denoted by $p^{\text{app}} = p^{\text{app}}(m_1, m_2, \ldots)$, is chosen. For the determination of favorable microstructure variables, a metric $d(p^a, p^*) \geq 0$ is chosen, in order measure the deviation from the prescribed properties. An optimization problem is solved for the microstructure variables

$$\min_{m_i \in M} d(p^a, p^*) \,.$$

3. *Uncertainties check.*

   In this post-processing step, for bounded properties, the obtained solution of the optimization problem is checked with the considered bounds to give information about maximal possible deviations for given microstructure moments. The incorporation of this step into the optimization problem would be highly cumbersome since, depending on imposed tolerances, the solution set might be too small or even empty.

4.2.2 Example 1: Linear elastic properties of orthotropic polycrystals of cubic single crystals

Overview of this example. In this example, a cubic material and polycrystals of it with orthotropic texture will be investigated regarding linear elastic properties. Based on the zeroth-order bounds, an artificial material database will be used for the selection of a possibly suitable material. Then, based on the zeroth-, first- and second-order bounds of this work, the properties-closure of the linear elastic properties of the textured polycrystals will be depicted to show the design possibilities offered by all orthotropic polycrystals of the chosen material. Finally, the ISC approach will be used as a pragmatic approach in order to determine textures delivering properties as close as possible to the desired properties-profile and the relative elastic range will be discussed. For the sake of simplicity, all physical quantities will be given without physical dimensions, and all numbers will be small.

1. Materials screening. In this example, the properties-profile given in Table 4.1 will be considered.

Table 4.1: Properties-profile of example 1

Maximum number of constituents:
2
Interesting quantities:
$\bar{C}_{1111} \approx 7.4$ with tolerances $\pm 3\%$
$\bar{C}_{1133} \approx -0.6$ with tolerances $\pm 3\%$
Macroscopic group:
Orthotropic with respect to $\{\boldsymbol{b}_i\}$

The material database tabulated in Table 4.2 is used in this example.

Table 4.2: Artificial material database with linear elastic properties

ID	Group	Properties	
M1	ISO	$\{\tilde{C}_{ijkl}\}$	$\left\{1,0,0,0,0,0,1,0,0,0,0,1,0,0,0,\frac{1}{2},0,0,\frac{1}{2},0,\frac{1}{2}\right\}$
M2	ISO	$\{\tilde{C}_{ijkl}\}$	$\left\{5,0,0,0,0,0,5,0,0,0,0,5,0,0,0,\frac{5}{2},0,0,\frac{5}{2},0,\frac{5}{2}\right\}$
M3	CUB	$\{\tilde{C}_{ijkl}\}$	$\left\{\frac{19}{3},\frac{4}{3},\frac{4}{3},0,0,0,\frac{19}{3},\frac{4}{3},0,0,0,\frac{19}{3},0,0,0,5,0,0,5,0,5\right\}$
M4	CUB	$\{\tilde{C}_{ijkl}\}$	$\left\{\frac{17}{3},\frac{5}{3},\frac{5}{3},0,0,0,\frac{17}{3},\frac{5}{3},0,0,0,\frac{17}{3},0,0,0,5,0,0,5,0,5\right\}$
M5	HEX	$\{\tilde{C}_{ijkl}\}$	$\{2,0,0,0,0,0,2,0,0,0,0,8,0,0,0,1,0,0,1,0,1\}$
M6	TRI	$\{\tilde{C}_{ijkl}\}$	$\left\{6,0,0,0,0,0,5,0,0,0,0,4,0,0,0,\frac{3}{2},0,0,1,0,\frac{1}{2}\right\}$

The material database Table 4.2 contains two isotropic materials (M1 and M2), two cubic materials (M3 and M4), one hexagonal / transversely isotropic material (M5) and one triclinic material (M6). The 21 free elastic constants are given as lists

$$\begin{aligned}\{\tilde{C}_{ijkl}\} \quad = \quad & \{\tilde{C}_{1111}, \tilde{C}_{1122}, \tilde{C}_{1133}, \tilde{C}_{1123}, \tilde{C}_{1113}, \tilde{C}_{1112}, \\ & \tilde{C}_{2222}, \tilde{C}_{2233}, \tilde{C}_{2223}, \tilde{C}_{1322}, \tilde{C}_{1222}, \\ & \tilde{C}_{3333}, \tilde{C}_{2333}, \tilde{C}_{1333}, \tilde{C}_{1233}, \\ & \tilde{C}_{2323}, \tilde{C}_{1323}, \tilde{C}_{1223}, \\ & \tilde{C}_{1313}, \tilde{C}_{1213}, \\ & \tilde{C}_{1212}\}\end{aligned} \tag{4.1}$$

for each material / SPPM. The single crystal stiffness tensor of the isotropic materials M1 and M2 can be expressed by the respective spectral decomposition as $\tilde{\mathbb{C}}_{M1} = 1\mathbb{P}_1 + 1\mathbb{P}_2$ and $\tilde{\mathbb{C}}_{M1} = 5\mathbb{P}_1 + 5\mathbb{P}_2$. The single crystal stiffness of the cubic materials M3 and M4 can be expressed by the respective spectral decomposition as $\tilde{\mathbb{C}}_{M3} = 9\mathbb{P}_1^{cub} + 5\mathbb{P}_2^{cub} + 10\mathbb{P}_3^{cub}$ and $\tilde{\mathbb{C}}_{M4} = 9\mathbb{P}_1^{cub} + 4\mathbb{P}_2^{cub} + 10\mathbb{P}_3^{cub}$. This tensorial expressions will facilitate the interpretation of the upcoming computations.

In order to see if a material or a combination of materials (MPPM) of Table 4.2 is already unable to deliver properties according to the properties-profile, the zeroth-order bounds $\mathbb{C}^{0\pm}$, see Eq. (3.59), can be computed. Before doing this, the difficulty anticipated in the introduction of $\tilde{\mathbb{C}}^{0+}$, see Eq. (3.61), is shortly discussed. The material database Table 4.2 contains 6 materials, i.e., $6 \cdot 5 = 30$ two phase material combinations exist, such that material designers would have to compute the zeroth-order bounds $\mathbb{C}^{0\pm}$ of each material and then of every two phase combination. As indicated in the introduction of $\tilde{\mathbb{C}}^{0+}$, the computation of the zeroth-order bound $\mathbb{C}^{0+}$ (and $\mathbb{C}^{0-}$ through the dual problem) for every new material combination is possible, but cumbersome and resource consuming for increasing number of materials in the database. Therefore, in order to save resources, the zeroth-order bounds $\tilde{\mathbb{C}}^{0\pm}$

should be evaluated for MPPMs. In this example, in order not to overload the reader, only the MPPMs M12, M34, M15 and M16 will be evaluated, together with the respective relative elastic volumes e_0, see Eq. (3.46). After that, the component bounds given in Eq. (3.44) are computed for the properties of interest. The results are tabulated in Table 4.3.

Table 4.3: Eigenvalues of zeroth-order bounds $\mathbb{C}^{0\pm}$ of each material and $\tilde{\mathbb{C}}^{0\pm}$ of selected MPPMs with corresponding relative elastic volume e_0 and bounds $\Gamma^{0\pm}_{1111}$ of $\bar{C}_{1111}$ and $\Gamma^{0\pm}_{1133}$ of $\bar{C}_{1133}$

ID	$\{\lambda^{0-}_{1,2}\}$	$\{\lambda^{0+}_{1,2}\}$	e_0	$\{\Gamma^{0\pm}_{1111}\}$	$\{\Gamma^{0\pm}_{1133}\}$
M1	$\{1.0, 1.0\}$	$\{1.0, 1.0\}$	1.0	$\{1.0, 1.0\}$	$\{0, 0\}$
M2	$\{5.0, 5.0\}$	$\{5.0, 5.0\}$	1.0	$\{5.0, 5.0\}$	$\{0, 0\}$
M3	$\{9.0, 5.0\}$	$\{9.0, 10.\}$	3.2×10^1	$\{6.3, 9.7\}$	$\{-1.2, 2.2\}$
M4	$\{9.0, 4.0\}$	$\{9.0, 10.\}$	9.8×10^1	$\{5.7, 9.7\}$	$\{-1.3, 2.7\}$
M5	$\{2.0, 2.0\}$	$\{10., 7.3\}$	3.3×10^3	$\{2.0, 8.3\}$	$\{-2.6, 3.7\}$
M6	$\{2.0, 2.0\}$	$\{7.0, 5.7\}$	6.5×10^2	$\{2.0, 6.1\}$	$\{-1.8, 2.3\}$
ID	$\{\tilde{\lambda}^{0-}_{1,2}\}$	$\{\tilde{\lambda}^{0+}_{1,2}\}$	e_0	$\{\Gamma^{0\pm}_{1111}\}$	$\{\Gamma^{0\pm}_{1133}\}$
M12	$\{1.0, 1.0\}$	$\{5.0, 5.0\}$	1.6×10^4	$\{1.0, 5.0\}$	$\{-2.0, 2.0\}$
M34	$\{9.0, 4.0\}$	$\{9.0, 10.\}$	9.8×10^1	$\{5.7, 9.7\}$	$\{-1.3, 2.7\}$
M15	$\{1.0, 1.0\}$	$\{10., 7.3\}$	2.1×10^5	$\{1.0, 8.3\}$	$\{-3.1, 4.2\}$
M16	$\{1.0, 1.0\}$	$\{7.0, 5.7\}$	4.1×10^4	$\{1.0, 6.1\}$	$\{-2.3, 2.8\}$

In Table 4.3 several properties of the MPPMs are visible. Let us first consider the SPPMs. Since M1 and M2 are isotropic materials, the zeroth-order bounds evaluate to the same values as in their spectral decompositions, as they should, since the effective stiffness equals the single crystal stiffness for such single-phase materials. The corresponding relative elastic volume equals identity, as it should. The zeroth-order bounds of the cubic materials evaluate to the expected results from literature according to their spectral decompositions, see, e.g., Lobos and Böhlke (2016). The respective relative elastic volume e_0 of the cubic materials is larger than the identity, but not by much, since the deviation from an isotropic state is not large, as seen in the eigenvalues of the corresponding zeroth-order bounds. The hexagonal material M5 shows higher intrinsic anisotropy, as reflected by the computed $\lambda^{\pm}_{1,2}$ and higher value of e_0. Finally, although the triclinic M6 does not deliver high $\lambda^{\pm}_{1,2}$ values as M5, it nevertheless shows intrinsic anisotropy comparable to M5, as seen by its e_0.

Now, let us interpret the results for the MPPMs. Although the materials M1 and M2 are relatively weak isotropic materials (compared to, e.g., M3 and M4), purposeful combination of a weak matrix with strategic positioning and arrangement of stiff inclusions can offer a huge anisotropic properties space. Exactly this intrinsic anisotropy is reflected by the corresponding zeroth-order bounds and the high relative elastic volume e_0 of M12 in Table 4.3. The high value of e_0 reflects only relative information, not absolute, as seen, e.g., in the smaller values of $\{\Gamma^{0\pm}_{1111}\}$ for M12 compared to M34. The MPPM M34, on the one hand, can offer high absolute values of $\{\Gamma^{0\pm}_{1111}\}$ but shows less flexibility, as indicated by the lower e_0. Analogous conclusions can be seen for M15 and M16.

Now, based on our properties-profile Table 4.1, M3 - M5, M34 and M15 are candidate MPPMs, since all others can be excluded by automatic evaluation due to the requirement $\bar{C}_{1111} \approx 7.4$. In order to be clear, no realization (independently of the microstructure) of M1, M2, M6, M12 or M16 exists, which is able to exceed its zeroth-order bounds and ever reach $\bar{C}_{1111} \approx 7.4$.

This automatic materials exclusion and search for candidates (in principle, since we still don't know if the prescribed properties are achievable) facilitates the materials selection/screening for large material databases. Several resources are also spared since unnecessary fabrications, tests or simulations of, e.g., MPPMs of M16 are avoided.

Naturally, the materials screening should cover not only requirements for the physical properties of a material, but chemical, manufacturing, financial and other restrictions. For this example, we will select an SPPM of the cubic material M4 and, according to the properties-profile Table 4.1, polycrystals of M4 with orthotropic texture will be investigated.

2. Microstructure optimization. The base material has been chosen. This allows us now to evaluate not only the zeroth- $\mathbb{C}^{0\pm}$, but also the first- $\mathbb{C}^{1\pm}$ and second-order bounds $\mathbb{C}^{2\pm}$ (based on HS, see Eq. (3.94)) of the present work.

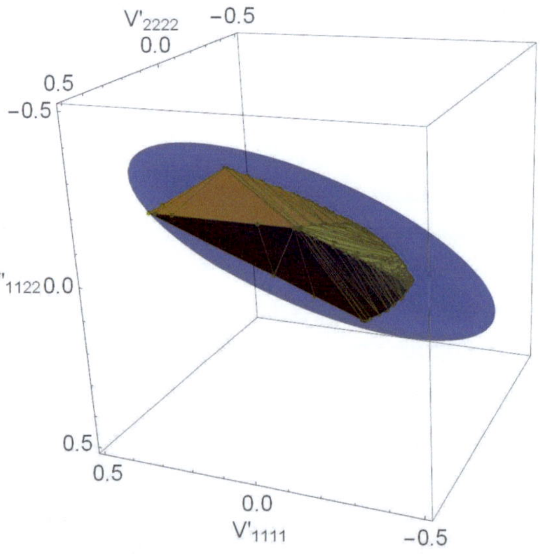

Figure 4.1: Regions $\check{\mathcal{V}}_4^{\text{ort}}$ and $\mathcal{N}_4$ depicted in yellow and blue, respectively.

The bounds $\mathbb{C}^{1\pm}$ and $\mathbb{C}^{2\pm}$ depend only on second- and fourth-order texture coefficients of the CODF of the cubic material M4. For cubic materials, all second-order reference tensors $\mathbb{T}'_{\langle 2\rangle\beta}$ vanish and only one fourth-order reference tensor $\mathbb{T}'_{\langle 4\rangle 1}$ exists, see Appendix A. This implies that the relevant influence of the CODF on polycrystals of M4 is embodied in the single fourth-order texture coefficient $\mathbb{V}'_{\langle 4\rangle 1} = \mathbb{V}'$. The polycrystals are restricted to be orthotropic. This restricts the number of degrees of freedom of $\mathbb{V}'$ to 3, V'_{1111}, V'_{1122} and V'_{2222} in this example, see Appendix A. In Fig. 4.1 the relevant regions $\check{\mathcal{V}}_4^{\text{ort}}$ (based on $\check{\mathcal{V}}^{\text{sym}}$, see Eq. (2.54), and constructed with S^{ort}, see Eq. (2.17), and Q_{15}, see Eq. (2.19)) and $\mathcal{N}_4$ are depicted in yellow and blue, respectively. It is shortly reminded that for cases with triclinic texture, the number of texture degrees of freedom increase and the delineation of the relevant texture set requires more resources. If the complexity is too high for certain applications, then the approximation of the CODF by CCODFs and the set of texture eigenvalues $\lambda_{2,4}$ defined in Eq. (2.39) and depicted in Fig. 2.1b offers an option.

The region $\check{\mathcal{V}}_4^{\text{ort}}$ depicted in Fig. 4.1 delivers a ground set for investigation and optimization of quantities depending on the orthotropic texture coefficient $\mathbb{V}'$. Each point in $\check{\mathcal{V}}_4^{\text{ort}}$ represents the influence of microstructures with orthotropic CODF in $\mathbb{C}^{1\pm}$ and $\mathbb{C}^{2\pm}$. Therefore, each point of $\check{\mathcal{V}}_4^{\text{ort}}$ delivers component bounds $\Gamma^{1/2\pm}_{1111/1133}$, see Eq. (3.44), which describe rectangles in the space of $\bar{C}_{1111}$ and $\bar{C}_{1133}$. Consider the regions depicted in Fig. 4.2.

In Fig. 4.2, the gray region depicts the zeroth-order bounds of M4 and corresponding component bounds Eq. (3.44). This is the zeroth-order properties-closure which reflects the material specific, microstructure independent assured limitations of a given material. The properties of the single crystal are marked by the black point on the left edge of the zeroth-order properties-closure. Now, for given material, evaluation a single point of $\check{\mathcal{V}}_4^{\text{ort}}$ together with the first-order bounds $\mathbb{C}^{1\pm}$ and corresponding component bounds Eq. (3.44) delivers the example light green rectangle depicted in Fig. 4.2. If the same point of $\check{\mathcal{V}}_4^{\text{ort}}$ is evaluated with the second-order bounds $\mathbb{C}^{2\pm}$, then the smaller red rectangle depicted in Fig. 4.2 is obtained, which is naturally contained in the one based on

$\mathbb{C}^{1\pm}$. Analogous evaluation of all points of $\check{\mathcal{V}}_4^{\mathrm{ort}}$ based on $\mathbb{C}^{1\pm}$ yields a set of rectangles, whose union delivers the first-order properties-closure, depicted by the green region in Fig. 4.2. The second-order properties-closure based on $\mathbb{C}^{2\pm}$ is depicted in red, while the region obtained with the ISC approach based on $\mathbb{C}^{ISC}$, see Eq. (3.115), is depicted in yellow in Fig. 4.2. The aimed properties of the properties-profile Table 4.1 are marked by the purple point and the purple rectangle (indicating the tolerances).

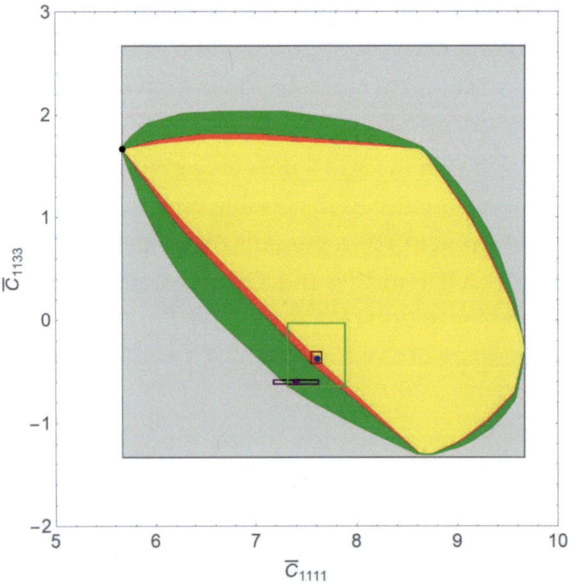

Figure 4.2: The single crystal properties of M4 are marked by the black point; the properties-closures of zeroth-, first- and second-order are represented by the gray, green and red regions, respectively; properties reached by the ISC approach are represented by the yellow region; the purple point and rectangle represent the aimed properties in the properties-profile of this example; the blue point represents the properties obtained by the ISC approach minimizing the euclidean distance to the purple point; the green and red rectangle represent the component bounds of first- and second-order for the optimized texture delivering the blue point of the ISC approach.

Material designers can observe several properties and difficulties in material design problems. The zeroth-order properties-closure gives primarily information of properties definitely out of reach of a material and only a notion of possibly achievable properties. The first-order properties-closure offers the same information, only taking into consideration volume fraction information. This means that the first-order properties-closure is able to show that some properties (initially considered as possibly achievable by the zeroth-order properties-closure) are impossible to reach for any polycrystal of the considered material. For properties-profile lying partly in the first-order properties-closure, it might be possible to find favorable microstructures. Evaluation of higher-order properties-closures, as the second-order one depicted in red in Fig. 4.2, might reveal that (under the assumed statistics) the aimed properties are definitely out of reach. Of course, the second-order bounds of the present work assume isotropic 2-point statistics and no long-range order, such that taking into consideration all possible 2-point statistics might help in this situation. But this is not assured. Therefore, it should be clear for every material designer that n-th-order bounds (and corresponding properties-closures) reflect properties limitation and offer only a notion of what might be achievable.

For cases as the present one, the only viable option (if a change of material is not possible) is to use a pragmatic approach. In this example, the ISC is used in order to determine texture coefficients delivering properties as close as possible to the properties-profile. The solution based on the ISC minimizing the Euclidean distance to the purple point is depicted by the blue point in Fig. 4.2.

If a change of material is allowed, then this should be attempted. The properties-closure of a new material with larger range of zeroth-order bounds and / or larger relative elastic might be able to cover the aimed properties. But again, higher-order bounds might show more information for certain problems.

3. Uncertainties check. Now, the corresponding optimal texture obtained with the ISC approach is used to evaluate again the first- and second-order properties-closure, which deliver the green and red rectangles in Fig. 4.2. This means that, based on $\mathbb{C}^{1\pm}$, the properties of polycrystals with the obtained optimal texture are assured to be in the green rectangle. If the polycrystals fulfill the HS statistical assumptions, then the properties are assured to lie in the remarkably small red rectangle, such that from a pragmatic point of view, the ISC approach delivers a reasonable estimate with small uncertainties for the present example.

As shortly remarked in Section 4.2.1, the incorporation of the prescribed tolerances of the properties in the optimization of the microstructure would have led in this example to an empty solution set, as already discussed and visible in Fig. 4.2. The afterward evaluation of the component bounds for the optimal texture based on the ISC approach offers the assured biggest possible deviations (green and red rectangles in Fig. 4.2) under assumed statistics. For closely related approaches taking into consideration all tolerances and gradients of the properties with respect to the texture, see, e.g., Lobos et al. (2015) or Lobos et al. (2017).

4.2.3 Example 2: Linear thermoelastic properties of a transversely isotropic matrix-inclusion composite

Overview of this example. In this example, the linear thermoelastic properties of a matrix-inclusion composite will be investigated. First, based on an artificial material database, the zeroth-order bounds will be used to select a material. Then, as in the previous example, the zeroth-, first- and second-order properties-closure will be depicted to show unreachable and possible achievable properties for the composite. This example is constructed to show possible applications for 2-step homogenization procedures, e.g., for fiber reinforced composites. In the first step, a single fiber might be enclosed in an isotropic material and homogenized, in order to obtain a transversely isotropic / hexagonal inclusion, which, in a second step, might be considered as a normal inclusion material in an isotropic matrix.

The materials screening and properties-closure presented in detail in Section 4.2.2 will also be used in this example.

1. Materials screening. We consider now the properties-profile tabulated in Table 4.4 and the material data given in Table 4.5, where the single crystal thermal stress coefficient components are given as lists

$$\{\tilde{\beta}_{ij}\} = \{\tilde{\beta}_{11}, \tilde{\beta}_{12}, \tilde{\beta}_{13}, \tilde{\beta}_{22}, \tilde{\beta}_{23}, \tilde{\beta}_{33}\} \ . \tag{4.2}$$

Table 4.4: Properties-profile of example 2

Maximum number of constituents:
2
Interesting quantities:
$\bar{C}_{1111} \approx 3$ with tolerances $\pm 10\%$
$\bar{\beta}_{33} \approx 4$ with tolerances $\pm 10\%$
Macroscopic group:
transversely isotropic with respect to $\boldsymbol{b}_3$

The materials given in Table 4.5 are the same ones as the ones given in Table 4.2 only extended by thermal properties, see Eq. (3.117).

The zeroth-order bounds defined in Eq. (3.70) are now computed for each material. For each of the 30 two-phase material combinations, the zeroth-order bounds $\underline{\underline{C}}^{0\pm}$, see Eq. (3.70), have to be calculated again and again. Instead of doing this, it is recommended to compute the zeroth-order bound $\underline{\underline{\tilde{C}}}_p^{0\pm}$ for each SPPM, see Eq. (3.72), and $\underline{\underline{\tilde{C}}}^{0\pm}$ for every MPPM, see Eq. (3.73). In this example, the bounds $\underline{\underline{\tilde{C}}}^{0\pm}$ will be computed (together with the corresponding component bounds of interest, see Appendix C or Eq. (3.58)), but only for the MPPMs of

Section 4.2.2, i.e., M12, M34, M15 and M16. The results are presented in Table 4.6.

Table 4.5: Material database for thermoelastic properties

ID	Group	Properties	
M1	ISO	$\{\tilde{C}_{ijkl}\}$	$\left\{1,0,0,0,0,0,1,0,0,0,0,1,0,0,0,\frac{1}{2},0,0,\frac{1}{2},0,\frac{1}{2}\right\}$
		$\{\tilde{\beta}_{ij}\}$	$\{1,0,0,1,0,1\}$
		$\tilde{f}$	1
M2	ISO	$\{\tilde{C}_{ijkl}\}$	$\left\{5,0,0,0,0,0,5,0,0,0,0,5,0,0,0,\frac{5}{2},0,0,\frac{5}{2},0,\frac{5}{2}\right\}$
		$\{\tilde{\beta}_{ij}\}$	$\{5,0,0,5,0,5\}$
		$\tilde{f}$	5
M3	CUB	$\{\tilde{C}_{ijkl}\}$	$\left\{\frac{19}{3},\frac{4}{3},\frac{4}{3},0,0,0,\frac{19}{3},\frac{4}{3},0,0,0,\frac{19}{3},0,0,0,5,0,0,5,0,5\right\}$
		$\{\tilde{\beta}_{ij}\}$	$\{5,0,0,5,0,5\}$
		$\tilde{f}$	5
M4	CUB	$\{\tilde{C}_{ijkl}\}$	$\left\{\frac{17}{3},\frac{5}{3},\frac{5}{3},0,0,0,\frac{17}{3},\frac{5}{3},0,0,0,\frac{17}{3},0,0,0,5,0,0,5,0,5\right\}$
		$\{\tilde{\beta}_{ij}\}$	$\{5,0,0,5,0,5\}$
		$\tilde{f}$	5
M5	HEX	$\{\tilde{C}_{ijkl}\}$	$\{2,0,0,0,0,0,2,0,0,0,0,8,0,0,0,1,0,0,1,0,1\}$
		$\{\tilde{\beta}_{ij}\}$	$\{2,0,0,2,0,8\}$
		$\tilde{f}$	5
M6	TRI	$\{\tilde{C}_{ijkl}\}$	$\left\{6,0,0,0,0,0,5,0,0,0,0,4,0,0,0,\frac{3}{2},0,0,1,0,\frac{1}{2}\right\}$
		$\{\tilde{\beta}_{ij}\}$	$\{6,0,0,5,0,4\}$
		$\tilde{f}$	6

The reader should notice that, compared to Table 4.2, e.g., for M1 already the bounds $\Gamma^{0\pm}_{1111}$ for $\bar{C}_{1111}$ show deviations from the isotropic single crystal behavior. This is due to the choice $s_0 = 0$ in the definition of the pragmatic bounds $\underline{\underline{\tilde{C}}}^{0\pm}$, see Eq. (3.72). This choice forces a discrepancy between the trivial bounds and induces inflation in all consequent component bounds. This non-optimality of the pragmatic bounds $\underline{\underline{\tilde{C}}}^{0\pm}$ is a clear disadvantage which can only be accepted. Nevertheless, the bounds $\underline{\underline{\tilde{C}}}^{0\pm}$ deliver valid bounds of the different materials, i.e., none of the materials can achieve properties outside the given values, such that automated materials screening based on these results is still possible. This is especially attractive for cases in which candidates for material combinations (MPPMs) for prescribed properties-profile are searched for in large databases. This task would be impossible to do with the bounds $\underline{\underline{C}}^{0\pm}$, which deliver tighter results but need the evaluation of the optimization problem for every new material combination. Based on the results given in Table 4.2, we are only able to exclude M1.

Table 4.6: Eigenvalues of zeroth-order bounds $\tilde{\underline{\underline{C}}}^{0\pm}$ of each material and of selected MPPMs with corresponding bounds $\Gamma^{0\pm}_{1111}$ of $\bar{C}_{1111}$ and $\gamma^{0\pm}_{37}$ of $-\bar{\beta}_{33}$

ID	$\{\tilde{\lambda}_1^{0-}, \tilde{\lambda}_2^{0-}, -\tilde{k}^{0-}\}$	$\{\tilde{\lambda}_1^{0+}, \tilde{\lambda}_2^{0+}, -\tilde{k}^{0+}\}$	$\{\Gamma^{0\pm}_{1111}\}$	$\{\gamma^{0\pm}_{37}\}$
M1	$\{0.37, 1., -5.73\}$	$\{2.73, 1., 0.73\}$	$\{0.79, 1.58\}$	$\{-1.13, 1.13\}$
M2	$\{0.52, 5., -21.73\}$	$\{13.66, 5., 3.66\}$	$\{3.51, 7.89\}$	$\{-5.27, 5.27\}$
M3	$\{0.93, 5., -14.3\}$	$\{17.66, 10., 3.66\}$	$\{3.64, 12.55\}$	$\{-6.32, 6.32\}$
M4	$\{0.93, 4., -14.3\}$	$\{17.66, 10., 3.66\}$	$\{2.98, 12.55\}$	$\{-6.56, 6.56\}$
M5	$\{0.45, 2., -18.73\}$	$\{17.25, 9.46, 12.88\}$	$\{1.48, 12.05\}$	$\{-9.14, 9.14\}$
M6	$\{0.45, 2., -19.73\}$	$\{14.28, 6.67, 6.73\}$	$\{1.48, 9.21\}$	$\{-7.15, 7.15\}$
ID	$\{\tilde{\lambda}_1^{0-}, \tilde{\lambda}_2^{0-}, -\tilde{k}^{0-}\}$	$\{\tilde{\lambda}_1^{0+}, \tilde{\lambda}_2^{0+}, -\tilde{k}^{0+}\}$	$\{\Gamma^{0\pm}_{1111}\}$	$\{\gamma^{0\pm}_{37}\}$
M12	$\{0.37, 1., -21.73\}$	$\{13.66, 5., 3.66\}$	$\{0.79, 7.89\}$	$\{-6.71, 6.71\}$
M34	$\{0.93, 4., -14.3\}$	$\{17.66, 10., 3.66\}$	$\{2.98, 12.55\}$	$\{-6.56, 6.56\}$
M15	$\{0.37, 1., -18.73\}$	$\{17.25, 9.46, 12.88\}$	$\{0.79, 12.05\}$	$\{-9.44, 9.44\}$
M16	$\{0.37, 1., -19.73\}$	$\{14.28, 6.67, 6.73\}$	$\{0.79, 9.21\}$	$\{-7.46, 7.46\}$

As mentioned in the previous example, materials screening contains a much broader collections of selection criteria. In this example, the 2-phase combination M15 is chosen, which might be considered as a light isotropic matrix with stiff transversely isotropic inclusions. For the MPPM M15, we now compute the zeroth-order bounds $\underline{\underline{C}}^{0\pm}$ to have a better insight concerning its limitations. The results are given in Table 4.7.

Table 4.7: Degrees of freedom of zeroth-order bounds $\underline{\underline{C}}^{0\pm}$ of M15 with corresponding bounds $\Gamma^{0\pm}_{1111}$ of $\bar{C}_{1111}$ and $\gamma^{0\pm}_{37}$ of $-\bar{\beta}_{33}$

ID	$\{\lambda_1^{0-}, \lambda_2^{0-}, -s^{0-}, -k^{0-}\}$	$\{\lambda_1^{0+}, \lambda_2^{0+}, -s^{0+}, -k^{0+}\}$	$\{\Gamma^{0\pm}_{1111}\}$	$\{\gamma^{0\pm}_{37}\}$
M15	$\{1., 1., -1., -14.\}$	$\{9.64, 8.44, -5.89, 7.31\}$	$\{1., 8.84\}$	$\{-9.91, 3.02\}$

The results given in Table 4.7 for M15 show clear deviations from the ones given in Table 4.6. This is due to the different definitions of $\underline{\underline{C}}^{0\pm}$ and $\tilde{\underline{\underline{C}}}^{0\pm}$. The valid statements of both bounds remains. The results of $\underline{\underline{C}}^{0\pm}$ are expected to reflect tighter bounds for some cases, due to their larger number of degrees of freedom.

2. Microstructure optimization. For the given materials, the first- and second-order bounds of this work can be computed, depending on the material concentrations and texture coefficients of second- and fourth-order.

The material M1 is isotropic, i.e., all corresponding reference tensors and texture coefficient vanish. The material M5 is hexagonal. For hexagonal materials only one second- and one fourth-order reference tensors exists, see Appendix A. This implies that the orientation averages for M5 depend solely on one second- and one fourth-order texture coefficient, denoted by $\mathbb{V}'_{\langle 2\rangle}$ and $\mathbb{V}'_{\langle 4\rangle}$. Since only hexagonal macroscopic behavior with respect to $\boldsymbol{b}_3$ is searched for, see Table 4.4, we are able to reduce the degrees of freedom of the texture. The texture is fully determined by V'_{11} (belonging to $\mathbb{V}'_{\langle 2\rangle}$) and V'_{1111} (belonging to $\mathbb{V}'_{\langle 4\rangle}$), see Appendix A. This implies that Eq. (2.55) describes the relevant texture set $\mathcal{V}_4^{\mathrm{hex}}$ for this example, depicted in Fig. 2.2.

It is shortly pointed out that, for the macroscopic hexagonal case of the present example, a CCODF could also be used, with underlying set for the texture eigenvalues (λ_2, λ_4), see Fig. 2.3. For the even more general cases with triclinic texture, the number of texture degrees of freedom of the texture coefficients increases, and the delineation of the relevant texture set requires more resources. If the complexity is too high for certain applications, then the approximation of the CODF by convex combination of CCODFs and the set of texture eigenvalues $\lambda_{2,4}$ defined in Eq. (2.39) and depicted in Fig. 2.1b offer an option.

Evaluation of the zeroth-, first- and second-order bounds and computation of the corresponding component bounds deliver the corresponding properties-closures depicted in Fig. 4.3.

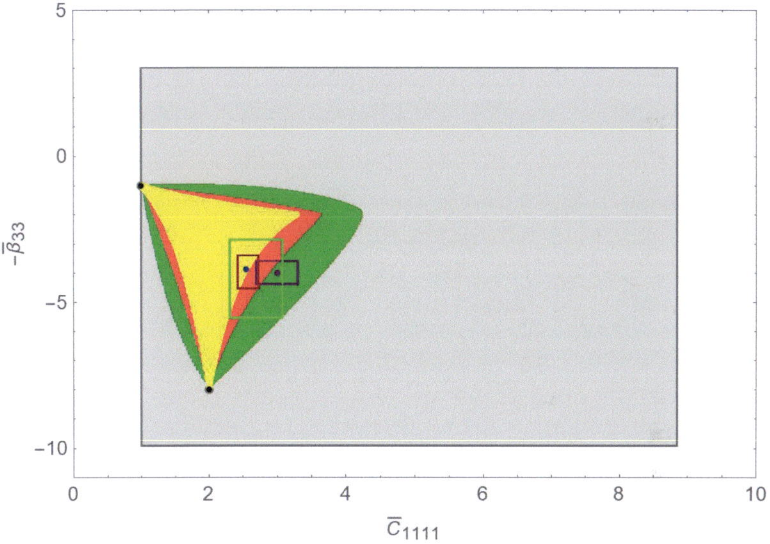

Figure 4.3: The single crystal properties of M1 and M5 are marked by the black points; the properties-closures of zeroth-, first- and second-order are represented by the gray, green and red regions, respectively; properties reached by the SA1 approach are represented by the yellow region; the purple point and rectangle represent the aimed properties in the properties-profile of this example; the blue point represents the properties obtained by the SA1 approach minimizing the euclidean distance to the purple point; the green and red rectangle represent the component bounds of first- and second-order for the optimized material concentration and texture delivering the blue point of the SA1 approach.

The zeroth-order properties-closure is depicted in Fig. 4.3 in gray, while the black points depict the single crystal properties of the materials

M1 and M5. The green and red region are the first- and second-order properties-closure, respectively, evaluated for all material concentrations and all texture points $(V'_{11}, V'_{1111}) \in \mathcal{V}_4^{\mathrm{hex}}$. The properties-closure is not necessarily a convex set, see the red region corresponding to the second-order properties-closure in Fig. 4.3. Further, it can be seen that the current first- and second-order properties-closures cover only a small portion of the zeroth-order properties-closure. This is mainly the implication of the restriction of hexagonal macroscopic behavior. Allowing for triclinic macroscopic behavior would allow for a wider range of textures and properties variations. Therefore, it should be noted that properties-profile restrictions might drastically reduce the space of linear thermoelastic properties possible reachable by selected materials.

The properties aimed at in the properties-profile Table 4.4 seem achievable by considering the first- and second-order bounds, as depicted by the purple point and rectangle in Fig. 4.3.

Now, we are interested in determining an optimal texture for the aimed properties. Since the MPPM of this example contains two constituents, the ISC approach is applicable only implicitly, making its computation slightly difficult. For a pragmatic and explicitly computable approach, the SA1 approximation, see Eq. (3.116), is considered in this example. Evaluation of all material concentrations and relevant texture points delivers the yellow region depicted in Fig. 4.3. Minimization of the Euclidean distance to the purple point (aimed properties) delivers the blue point in Fig. 4.3.

3. Uncertainties check. The optimized material concentrations and texture are now used to evaluate the first- and second-order bounds of the present work. The corresponding component bounds are depicted in Fig. 4.3 by the green (first-order) and red (second-order) rectangles, respectively. It can be seen that, based on the first-order bounds, for the optimized material concentrations and texture, a fair amount of the aimed properties might be achievable. Based on the second-order bounds of this work, the optimized statistics seem to yield an almost negligible intersection with the aimed properties range, with respect to $\bar{C}_{1111}$, but an acceptable window with respect to $\bar{\beta}_{33}$. This is the case due to optimization only of the Euclidean distance from the purple point denoting the aimed properties. As indicated at the end of the previous example, approaches considering the tolerances and gradients of properties of interest can be taken into account, as done in Lobos et al. (2015) or Lobos et al. (2017). But, as clearly visible through the shape of the yellow region of the SA1 approach in Fig. 4.3, the reader should realize that for this and surely other cases, several approximations of the exact material behavior will not be able to deliver satisfying results for all properties with corresponding tolerances.

4.3 Bounds of nonlinear properties

4.3.1 Objectives

In this section, we are interested in the applications of the texture dependent expressions of the present work in the context of nonlinear behavior. Therefore, the nonlinear bounds briefly sketched in Section 3.6 will be computed for viscoplastic single-phase polycrystals with texture. The aim of this example is not to compute approximations of the nonlinear behavior but mainly to illustrate the application of the texture dependent expressions of this work. Further, differences concerning the texture dependency of the approach of this work with the approach of DeBotton and Ponte Castañeda (1995) for viscoplastic polycrystals build upon Ponte Castañeda (1991) are discussed for clarity. These texture-dependency differences are also important for approaches extending DeBotton and Ponte Castañeda (1995), see, e.g., Ponte Castañeda (2002), Liu and Ponte Castañeda (2004) and Ponte Castañeda (2015) for improved bounds and estimates for known texture. At the end of this example, for cubic materials with cubic texture, the effective flow stress will be bounded by expressions computable by hand.

4.3.2 Example 3: Viscoplastic properties of textured polycrystalline materials

Rigid-viscoplastic material behavior. A rigid-viscoplastic material behavior is referred to as a material law in which the stress is governed not by the applied deformation but by the rate of deformation. Such material laws are not able to reproduce spring-back phenomena, but provide examples for which homogenization theory can be applied. Consider a rigid-viscoplastic material law with strain rate potential U, i.e. $\boldsymbol{\varepsilon} = \partial U / \partial \boldsymbol{\sigma}$ ($\boldsymbol{\varepsilon}$ denoting the strain rate in this section, i.e, $\boldsymbol{\varepsilon}$ represents the symmetrized spatial gradient of the velocity field), in an SPPM. The strain rate potential is defined in this example as the sum of all potentials of the n_s slip systems, i.e., $U = \sum_{k=1}^{n_s} \phi_k(\tau_k)$, where ϕ_k are convex functions of the shear stress τ_k in the k-th slip system. The convex functions are assumed to have a growth higher than quadratic, i.e., $\phi_k(\tau) > O(\tau^2)$. At a material point $\boldsymbol{x}$, the shear stress in the corresponding slip system is described as $\tau_k(\boldsymbol{x}) = \boldsymbol{\sigma}(\boldsymbol{x}) \cdot \boldsymbol{\mu}(\boldsymbol{x})$, with $\boldsymbol{\mu}(\boldsymbol{x}) = \boldsymbol{Q}(\boldsymbol{x}) \star \tilde{\boldsymbol{\mu}}_k$. The reference Schmidt-tensors $\tilde{\boldsymbol{\mu}}_k$ are defined as the symmetrization of the slip directions $\tilde{\boldsymbol{d}}_k$ and slip plane normals $\tilde{\boldsymbol{n}}_k$ of the corresponding k-th slip system, i.e., $\tilde{\boldsymbol{\mu}}_k = \frac{1}{2}(\tilde{\boldsymbol{n}}_k \otimes \tilde{\boldsymbol{d}}_k + \tilde{\boldsymbol{d}}_k \otimes \tilde{\boldsymbol{n}}_k)$. The orientation tensor $\boldsymbol{Q}(\boldsymbol{x})$ reorients the reference crystallographic axis at the momentary position. In this example we consider the standard power law, see, e.g., Hutchinson (1976)

$$\phi_k(\tau) = \frac{\gamma_{0k}\tau_{0k}}{n+1} \left(\frac{\tau}{\tau_{0k}} \right)^{n+1} \tag{4.3}$$

for odd $n \geq 3$ in this example with the constant material specific reference shear stress τ_{0k} and the reference shear rate γ_{0k} of the k-th slip system. The material parameter n is bounded in general as $n \in [1, \infty)$, where the limits $n \to 1$ and $n \to \infty$ correspond to linear viscous and ideally plastic material behavior, respectively. The material law (flow rule) evaluates to

$$\varepsilon = \frac{\partial U}{\partial \boldsymbol{\sigma}} = \sum_{k=1}^{n_s} \gamma_{0k} \left(\frac{\tau_k}{\tau_{0k}} \right)^n \boldsymbol{\mu}_k(\boldsymbol{x}) \,. \tag{4.4}$$

For simplicity, from here on we reformulate Eq. (4.3) as

$$\phi_k(\tau) = \frac{1}{n+1} \lambda_k \tau^{n+1} \,, \quad \lambda_k = \frac{\gamma_{0k}}{\tau_{0k}^n} > 0 \,. \tag{4.5}$$

As illustrated in Section 3.6, the effective potential is defined as

$$\bar{U} = \inf_{\boldsymbol{\sigma} \in S} \langle U \rangle_v \,, \quad S = \{ \boldsymbol{\sigma} \mid \operatorname{div}(\boldsymbol{\sigma}) = \boldsymbol{o} \wedge \langle \boldsymbol{\sigma} \rangle_v = \bar{\boldsymbol{\sigma}} \} \,, \tag{4.6}$$

and the macroscopic strain rate is obtained through

$$\bar{\varepsilon} = \frac{\partial \bar{U}}{\partial \bar{\boldsymbol{\sigma}}} \,. \tag{4.7}$$

Now, $\gamma_{0k} = \gamma_0 \forall k$ is chosen, in order to simplify some expressions. Since n is the same for all slip systems and all grains in the polycrystal, the effective potential is a homogeneous function of degree $(n+1)$ in $\bar{\boldsymbol{\sigma}}$. The effective potential $\bar{U}$ depends on $\bar{\boldsymbol{\sigma}}$ through its deviatoric part $\bar{\boldsymbol{\sigma}}' = \bar{\boldsymbol{\sigma}} - \frac{1}{3}\operatorname{tr}(\bar{\boldsymbol{\sigma}})\, \boldsymbol{I}$. Based on the (von Mises equivalent) stress $\bar{\sigma}_e = \sqrt{\frac{3}{2}\bar{\boldsymbol{\sigma}}' \cdot \bar{\boldsymbol{\sigma}}'}$, the effective potential $\bar{U}$ can be expressed as

$$\bar{U} = \frac{\gamma_0 \bar{\sigma}_f}{n+1} \left(\frac{\bar{\sigma}_e}{\bar{\sigma}_f} \right)^{n+1} \,, \quad \bar{\sigma}_f = \left(\frac{\gamma_0 \bar{\sigma}_e^{n+1}}{(n+1)\bar{U}} \right)^{\frac{1}{n}} \,, \tag{4.8}$$

where $\bar{\sigma}_f$ denotes the effective flow stress of the polycrystal, which depends, in general, on the loading and the microstructure of the polycrystal, see, e.g., Nebozhyn et al. (2001). For untextured polycrystals further simplifications are possible, see Dendievel et al. (1991). If a lower bound of $\bar{U}$ is obtained, then an upper bound for $\bar{\sigma}_f$ can be computed.

The LCC. Now, following the procedure described in Section 3.6 based on the comparison with linear material laws and corresponding quadratic potentials, we introduce the comparison potential of the LCC in the k-th slip system

$$\phi_{kc}(\tau) = \frac{1}{2}\lambda_{kc}\tau^2 + e_{kc}\tau + \frac{1}{2}l_{kc} , \quad \lambda_{kc} \geq 0 , \tag{4.9}$$

and the quantities

$$\begin{aligned} \varphi_{kc} &= \sup_{\tau}\{\phi_{kc}(\tau) - \phi_k(\tau)\} \geq \phi_{kc}(\tau) - \phi_k(\tau)\ \forall\tau , \\ \tau_{kc} &= \arg\sup_{\tau}\{\phi_{kc}(\tau) - \phi_k(\tau)\} , \end{aligned} \tag{4.10}$$

for each slip system. It should be noted that the quantity φ_{kc} is stationary with respect to τ_{kc} and depends in this example solely on the constant material parameters λ_k and the variables λ_{kc}, e_{kc} and l_{kc} of the LCC potential. For example, consider the case $e_{kc} = 0$, then $\tau_{kc} = \sqrt[n-1]{(\lambda_{kc}/\lambda_k)}$ holds.

Now, based on φ_{kc}, we can bound the effective potential as follows

$$\frac{1}{v}\int_v \sum_{k=1}^{n_s} \phi_k(\tau_k)\mathrm{d}v \geq \frac{1}{v}\int_v \sum_{k=1}^{n_s}[\phi_{kc}(\tau_k) - \varphi_{kc}]\mathrm{d}v . \tag{4.11}$$

Taking the infimum of each side delivers

$$\bar{U} \geq \bar{U}_c - \bar{V}_c , \quad \bar{V}_c = \left\langle \sum_k^{n_s} \varphi_{kc} \right\rangle_v = \sum_k^{n_s} \varphi_{kc} , \tag{4.12}$$

and the effective potential of the linear comparison material

$$\begin{aligned} \bar{U}_c &= \inf_{\boldsymbol{\sigma} \in S} \frac{1}{v} \int_v \sum_{k=1}^{n_s} \phi_{kc}(\tau_k) \mathrm{d}v \\ &= \inf_{\boldsymbol{\sigma} \in S} \frac{1}{v} \int_v \frac{1}{2} \boldsymbol{\sigma}(\boldsymbol{x}) \cdot \mathbb{S}(\boldsymbol{x})[\boldsymbol{\sigma}(\boldsymbol{x})] + \boldsymbol{\sigma}(\boldsymbol{x}) \cdot \boldsymbol{e}(\boldsymbol{x}) + \frac{1}{2} l(\boldsymbol{x}) \mathrm{d}v \end{aligned} \tag{4.13}$$

with

$$\begin{aligned} \mathbb{S}(\boldsymbol{x}) = \boldsymbol{Q}(\boldsymbol{x}) \star \tilde{\mathbb{S}} \,, \quad & \tilde{\mathbb{S}} = \sum_{k=1}^{n_s} \lambda_{kc} \tilde{\boldsymbol{\mu}}_k^{\otimes 2} \,, \\ \boldsymbol{e}(\boldsymbol{x}) = \boldsymbol{Q}(\boldsymbol{x}) \star \tilde{\boldsymbol{e}} \,, \quad & \tilde{\boldsymbol{e}} = \sum_{k=1}^{n_s} e_{kc} \tilde{\boldsymbol{\mu}}_k \,, \\ l(\boldsymbol{x}) = \tilde{l} = & \sum_{k=1}^{n_s} l_{kc} \,. \end{aligned} \tag{4.14}$$

The reader should note that, for the present example, $\bar{U}_c$ represents the effective potential of a "standard linear SPPM with eigenfields". The corresponding CODF of the slip system can be treated as $f(\boldsymbol{Q})$ in the preceding examples. Naturally, the present approach can be extended with small modifications to MPPMs, see, e.g., Ponte Castañeda (1991), Ponte Castañeda and Suquet (1997).

Bounds of HS type in terms of texture coefficients. The lower bound of $\bar{U}$ obtained in Eq. (4.12) is valid independently of the choice of $\lambda_{kc} \geq 0$, e_{kc} and l_{kc}. This freedom would allow to optimize the lower bound, for given $\bar{U}_c$. Since $\bar{U}_c$ is not known, we use the lower HS bound U_c^{HS} of $\bar{U}_c$, for which $\bar{U} \geq \bar{U}_c - \bar{V}_c \geq U_c^{HS} - \bar{V}_c = B^{HS}$ holds for arbitrary LCC parameters and comparison compliance $\mathbb{S}_0$ fulfilling $\mathbb{S}_0 \leq \mathbb{S}_c$. In this work, isotropic $\mathbb{S}_0$ and isotropic two-point statistics in the polycrystalline phases are assumed, such that the HS bound are immediately obtained in terms of texture coefficients of second- and fourth-order.

At this point, the author would like to shortly point out the difference of the approach of this work with the approach of DeBotton and Ponte Castañeda (1995). Based on the original variational principle of Ponte Castañeda (1991), DeBotton and Ponte Castañeda (1995) compute a lower bound of $\bar{U}$ by considering a finite number of given crystallographic orientations and using a different compliance tensor for each orientation, considered as a distinct phase. This allows DeBotton and Ponte Castañeda (1995) to optimize their bounds further, since this choice has more degrees of freedom than the approach of this work, resulting in a bound nearer to $\bar{U}$ than the one of this work. The drawback of DeBotton and Ponte Castañeda (1995) is that due to the change of compliance from orientation to orientation, the average over all orientations cannot be treated only based on the CODF of the slip system as in the present work. The approach of the present work is a special case of DeBotton and Ponte Castañeda (1995) (for vanishing e_{ck} and l_{ck}) which considers a single compliance for all orientations. This is an implication of the approach based on Eq. (4.9). The author took this approach to derive expressions treatable as a standard polycrystal and, therefore, expressible in terms of texture coefficients. The trade-off between DeBotton and Ponte Castañeda (1995) and the present approach is a higher accuracy of the resulting bounds vs. a practical parametrization for materials design. For given texture in terms of single crystal orientations, the approach of DeBotton and Ponte Castañeda (1995) may be computed for results with better accuracy of the resulting bounds than the present approach. But, if the texture contains a large number of orientations, is not known or to be varied arbitrarily for materials design objectives, the approach of DeBotton and Ponte Castañeda (1995) is not practical. The approach of this work allows investigating the influence of texture by variation of the second- and fourth-order texture coefficients, without any specification of the CODF. This has the cost of less accuracy of the resulting bounds, in the sense that the resulting bounds are nevertheless valid, but farther from the effective $\bar{U}$. These texture-dependency differences apply to extended approaches as Ponte Castañeda (2002), Liu and Ponte Castañeda (2004) and Ponte Castañeda (2015), where the LCC is used for bounds and estimates not only based on compliances, but also with eigen-strain-rates, but with given texture.

We return now to the approach of the present work. Optimization of the lower bound B^{HS} delivers

$$\begin{aligned} \bar{U} \geq B_s^{HS} &= \sup_{\rho \in \mathcal{C}} B^{HS} = B^{HS}|_{\rho_s} \,, \\ \rho &= \{\lambda_{kc}, e_{kc}, l_{kc}, \tau_{kc}, \mathbb{S}_0\} \,, \\ \mathcal{C} &= \{\rho \mid \lambda_{kc} \geq 0 \wedge \tau_{kc} = \arg\sup_\tau \{\phi_{kc} - \phi_k\} \wedge \mathbb{S}_0 \leq \mathbb{S}\} \end{aligned} \tag{4.15}$$

where the parameters ρ_s deliver the supremum. These parameters ρ_s depend, in general, on $\bar{\boldsymbol{\sigma}}$ *and on the crystallographic texture*, i.e., in this example, on the second- and fourth-order texture coefficients of the CODF.

It should be noted, that due to the dependency of ρ_s on $\bar{\boldsymbol{\sigma}}$ (and on the texture coefficients), the quantity B_s^{HS} is highly nonlinear in $\bar{\boldsymbol{\sigma}}$. The gradient of B_s^{HS} with respect to $\bar{\boldsymbol{\sigma}}$ (for the examination of estimates of the effective strain rate) is cumbersome to compute if the chain rule is to be considered, i.e.,

$$\frac{\mathrm{d}B_s^{HS}}{\mathrm{d}\bar{\boldsymbol{\sigma}}} = \frac{\partial B_s^{HS}}{\partial \bar{\boldsymbol{\sigma}}} + \frac{\partial B_s^{HS}}{\partial \rho_s} \frac{\partial \rho_s}{\partial \bar{\boldsymbol{\sigma}}} \,. \tag{4.16}$$

But, it should be noted, that only 2 cases are to be considered. The first one is the one where the supremum of Eq. (4.15) is obtained at an inner point of the constraints set $\mathcal{C}$. If this is the case, then B_s^{HS} is stationary with respect to ρ_s and the terms of chain rule vanish immediately. This implies that the gradient of B_s^{HS} with respect to $\bar{\boldsymbol{\sigma}}$ is given solely by the partial derivative of B_s^{HS} with respect to $\bar{\boldsymbol{\sigma}}$ for fixed ρ_s. The second case is where the supremum is achieved at a boundary point of $\mathcal{C}$. If the constraints which describe $\mathcal{C}$ do not depend on $\bar{\boldsymbol{\sigma}}$ which is the case in this optimization problem, then the value of ρ_s is independent of $\bar{\boldsymbol{\sigma}}$. This implies that the chain rule term also vanishes in this second case. Therefore, these cases imply that the effective strain rate estimate $\bar{\varepsilon}^{HS}$ is obtained, for the present example, by the partial derivative $\partial B_s^{HS}/\partial\bar{\boldsymbol{\sigma}}$ evaluated at ρ_s, i.e.,

$$\bar{\varepsilon}^{HS} = \frac{\partial B_s^{HS}}{\partial \bar{\boldsymbol{\sigma}}} = \frac{\partial U_c^{HS}}{\partial \bar{\boldsymbol{\sigma}}}\bigg|_{\rho=\rho_s} = \mathbb{S}^{HS}[\bar{\boldsymbol{\sigma}}] + \boldsymbol{e}^{HS}\bigg|_{\rho=\rho_s}. \tag{4.17}$$

The effective stress-strain-rate law is nevertheless nonlinear, since ρ_s depends on $\bar{\boldsymbol{\sigma}}$. The relation Eq. (4.17) extends to the local fields, such that these fields might be localized with the localization relations of the HS results.

From the perspective of materials design, based on the material law Eq. (4.17), the strain rates might be investigated in terms of the texture coefficients for interesting applications searching for tendencies due to texture changes.

Analytic example: simplified polycrystal of FCC crystals with cubic texture. For a face centered cubic (FCC) material, the slip system is described with the octahedral system containing $n_s = 12$ slip systems, see, e.g., Böhlke (2001). In this computational example, $\tau_{0k} = \tau_0$, $\lambda_k = \gamma_0/\tau_0^n \forall k$ are chosen in order to discuss the dimensionless ratio of the effective flow stress $\bar{\sigma}_f$ and τ_0.

The LCC might be simplified in several ways. If all λ_{kc} are chosen to be different, the single crystal reference compliance $\tilde{\mathbb{S}} = \sum_{k=1}^{12} \lambda_{kc}\tilde{\boldsymbol{\mu}}_k$ of the FCC material is not cubic but triclinic. In order to present an example computable by hand, $\lambda_{kc} = \frac{\gamma_0}{\tau_0}\lambda_c$, $e_{kc} = e_c = 0$, $l_{kc} = l_c \forall k$ are chosen. Under this simplification, $\tilde{\mathbb{S}}$ is cubic, $\tilde{e}$ vanishes and $\tilde{l} = 12 l_c$. The single crystal compliance is expressible in its cubic spectral decomposition $\tilde{\mathbb{S}} = \frac{\gamma_0}{\tau_0}\sum_{\alpha=1}^{3} \lambda_\alpha^{cub}\mathbb{P}_\alpha^{cub}$ with dimensionless $\{\lambda_\alpha^{cub}\} = \{0, 2\lambda_c, 2\lambda_c/3\}$, see Appendix A, Eq. (A.12).

For a polycrystal of a FCC material with cubic texture, the HS complementary potential U^{HS} with cubic $\mathbb{S}^{HS} = \frac{\gamma_0}{\tau_0}\sum_{\alpha=1}^{3}\zeta_\alpha^{HS}\mathbb{P}_\alpha^{cub}$ based on an isotropic comparison compliance $\mathbb{S}_0 = \frac{\gamma_0}{\tau_0}(s_1\mathbb{P}_1 + s_2\mathbb{P}_2) \leq \mathbb{S}$ is simplified. The dimensionless eigenvalues of the cubic $\mathbb{S}^{HS}$ evaluate to

$$\begin{aligned} \zeta_1^{HS} &= 0\,, \\ \zeta_2^{HS} &= -\frac{2\lambda_c(5\lambda_c(4s_1+3s_2)+3s_2(2V_1'+3)(3s_1+s_2))}{\lambda_c(6V_1'-11)(4s_1+3s_2)-15s_2(3s_1+s_2)}\,, \\ \zeta_3^{HS} &= \frac{2\lambda_c(5\lambda_c(4s_1+3s_2)+s_2(9-4V_1')(3s_1+s_2))}{\lambda_c(4V_1'+11)(4s_1+3s_2)+15s_2(3s_1+s_2)} \end{aligned} \tag{4.18}$$

where only ζ_2^{HS} and ζ_3^{HS} depend on the single degree of freedom $V_{1111}' = \sqrt{2/15}V_1'$, $V_1' \in [-13/27, 1]$ (see Lobos et al. (2017)) of the cubic fourth-order texture coefficient $\mathbb{V}_{\langle 4\rangle}'$ of the cubic $\tilde{\mathbb{S}}$. For the fulfillment of $\mathbb{S}_0 \leq \mathbb{S}$, $s_1 = 0$ is chosen and s_2 is constrained as $0 \leq s_2 \leq 2\lambda_c/3$. For a tensile test with effective stress $\bar{\boldsymbol{\sigma}} = \bar{\sigma}_{11}\boldsymbol{b}_1^{\otimes 2}$, the expression U^{HS} can be obtained by long but simple computations as

$$U_c^{HS} = \gamma_0\tau_0\frac{1}{3}\zeta_2^{HS}\left(\frac{\bar{\sigma}_{11}}{\tau_0}\right)^2 + 6l_c\,. \tag{4.19}$$

With $\tau_{kc} = \tau_0\sqrt[n-1]{\lambda_c}$, the quantities φ_{kc} simplify to

$$\varphi_{kc} = \phi_{kc}(\tau_{kc}) - \phi_k(\tau_{kc}) = \gamma_0\tau_0\frac{1}{2}\frac{n-1}{n+1}\lambda_c^{\frac{n+1}{n-1}} + \frac{1}{2}l_c \tag{4.20}$$

such that $\bar{V}_c$ evaluates to

$$\bar{V}_c = \gamma_0\tau_0 6\frac{n-1}{n+1}\lambda_c^{\frac{n+1}{n-1}} + 6l_c\,. \tag{4.21}$$

The lower bound B^{HS} is given by

$$B^{HS} = \gamma_0\tau_0\hat{B}^{HS}\,, \quad \hat{B}^{HS}\frac{1}{3}\zeta_2^{HS}\left(\frac{\bar{\sigma}_{11}}{\tau_0}\right)^2 - 6\frac{n-1}{n+1}\lambda_c^{\frac{n+1}{n-1}} \tag{4.22}$$

having no dependency on l_c. The best possible dimensionless bound $\hat{B}^{HS}$ is obtained by solving the optimization problem with dimensionless parameters λ_c and s_2

$$B_s^{HS} = \gamma_0 \tau_0 \hat{B}_s^{HS} \,, \quad \hat{B}_s^{HS} = \hat{B}^{HS}|_{(\lambda_c, s_2)=\rho_s} \,,$$
$$\rho_s = \arg \sup_{(\lambda_c, s_2)\in\mathcal{C}} \hat{B}^{HS} \,, \tag{4.23}$$
$$\mathcal{C} = \{(\lambda_c, s_2) \,|\, 0 \leq \lambda_c \wedge 0 \leq s_2 \leq \tfrac{2}{3}\lambda_c\} \,.$$

The solution of the dimensionless optimization problem is found to evaluate to

$$\rho_s = (\lambda_{sc}, s_{s2}) \,,$$
$$\lambda_{sc} = 3^{(1-n)} \left(\frac{\bar{\sigma}_{11}}{\tau_0}\right)^{n-1} \left(\frac{4V_1' + 21}{43 - 18V_1'}\right)^{\frac{n-1}{2}} \,, \quad s_{s2} = \frac{2}{3}\lambda_{sc} \,. \tag{4.24}$$

The dimensionless potential evaluates to

$$\hat{B}_s^{HS} = \frac{4 \left(\frac{4V_1'+21}{43-18V_1'}\right)^{\frac{n+1}{2}}}{3^n (n+1)} \left(\frac{\bar{\sigma}_{11}}{\tau_0}\right)^{n+1} \tag{4.25}$$

It should be noted, that these analytic results are, as in the general case, $\bar{\boldsymbol{\sigma}}$ *and texture dependent*. The solution of the optimization problem can be used in order to generate the HS estimate of the normalized strain rate

$$\frac{\bar{\boldsymbol{\varepsilon}}^{HS}}{\gamma_0} = \left(\sum_{\alpha=1}^{3} \zeta_\alpha^{HS} \mathbb{P}_\alpha^{cub}\right) \left[\frac{\bar{\boldsymbol{\sigma}}}{\tau_0}\right] \Big|_{(\lambda_c, s_2)=\rho_s} \,. \tag{4.26}$$

The HS bound for the dimensionless effective flow stress $\bar{\sigma}_f/\tau_0 \leq \bar{\sigma}^{HS}/\tau_0$ is described following Eq. (4.8) with $\bar{\sigma}_e = \bar{\sigma}_{11}$ as

$$\frac{\bar{\sigma}_f^{HS}}{\tau_0} = \frac{3}{4^{\frac{1}{n}}} \left(\frac{4V_1' + 21}{43 - 18V_1'}\right)^{-\frac{n+1}{2n}} \geq \frac{\bar{\sigma}_f}{\tau_0} \,, \tag{4.27}$$

which offers a loading independent, texture dependent compact expression. The limit for $n \to \infty$ yields

$$\lim_{n\to\infty} \frac{\bar{\sigma}_f^{HS}}{\tau_0} = 3\sqrt{\frac{43-18V_1'}{4V_1'+21}}\,. \tag{4.28}$$

First, results are presented now for $n = 10$, $\bar{\sigma}_{11}/\tau_0 \in [0,6]$ and varying texture, see Fig. 4.4.

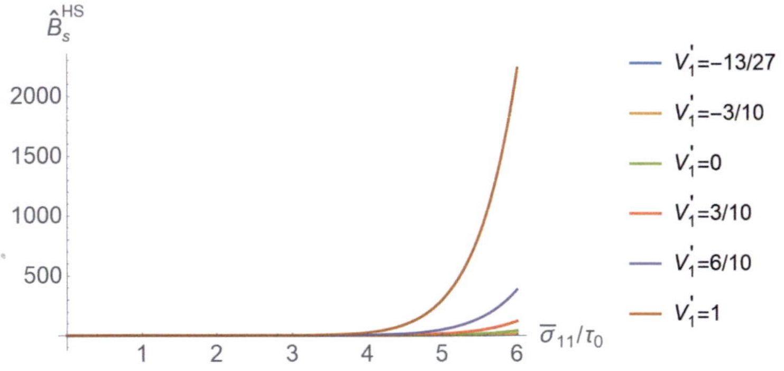

(a)

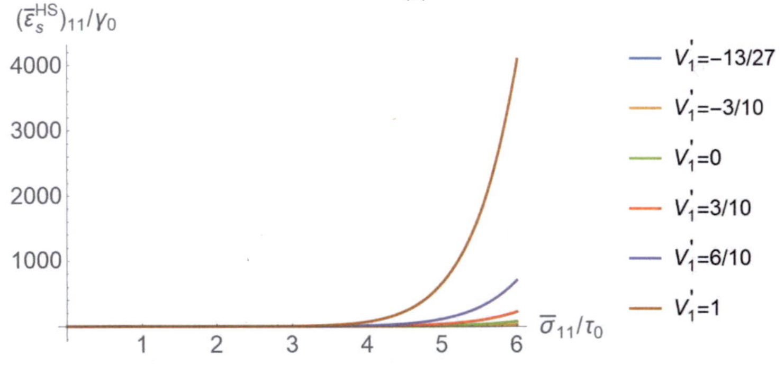

(b)

Figure 4.4: (a) Effective dimensionless lower bound $\hat{B}_s^{HS}$ based on HS (with $n = 10$); (b) Resulting nonlinear effective stress - strain rate relation estimate based on $\hat{B}_s^{HS}$ (with $n = 10$)

In Fig. 4.4a the dimensionless effective HS potential is depicted in terms of $\bar{\sigma}_{11}/\tau_0$. It is visible that depending on the fixed texture, the gradient with respect to the effective stress changes drastically, which naturally affects the dimensionless effective nonlinear material law depicted in Fig. 4.4b. This motivates the tendencies of strain rates changes depending on the texture and corresponding texture coefficients of second- and fourth-order. These tendencies can be analyzed based on the convex set of texture coefficients.

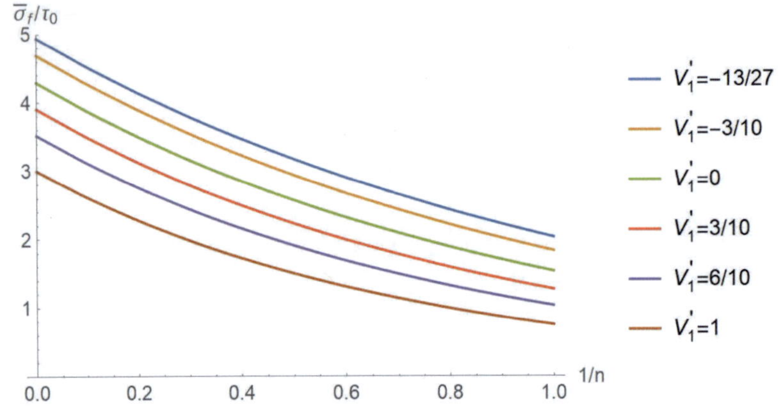

Figure 4.5: Effective flow stress estimated based on HS against $1/n$ for varying texture

In Fig. 4.5 the dimensionless effective flow stress bound $\bar{\sigma}^{HS}/\tau_0$ is represented for varying texture. The bound gives information of texture dependent limitations of the effective flow stress. The tendencies for varying cubic texture are visible, for $V_1' \to -13/27$ the bound is increased, while in the opposite direction it is decreased independently of the specific viscous exponent n. For materials design purposes, the texture dependent bound can help designers, e.g., to exclude materials which definitely will not be able to achieve certain desired effective flow stresses. Also interesting are the limits $n \to \infty$ or $1/n \to 0$ since these limits give information for material designers about the texture dependent bounds of ideally plastic material behavior.

It is shortly remarked that the derived relations of this section depending on V_1' are also valid for a single CCODF, which preserves the microscopic cubic behavior, as derived and pointed out for Eq. (2.76). As discussed in Lobos et al. (2017) for elasticity expressions, V_1' can immediately be replaced with the texture eigenvalue λ_4 of a CCODF, fulfilling the bounds Eq. (2.37) depicted in Fig. 2.1b.

Finally, the author would like to stress, that the present relations for the effective flow stress offer only upper bounds, no accurate approximations. As discussed in this section, for a higher bound accuracy, the approach of DeBotton and Ponte Castañeda (1995) may be considered, but at the cost of higher computation complexity and less flexibility concerning varying texture. The texture dependent bounds of this example are extremely compact and fairly easily derivable by the texture dependent expressions of the present work. The reader should notice, that other bounds known from literature, for example, the corresponding Voigt (constant deformation) or Reuss (constant stress) bounds for nonlinear behavior could be computed, see, e.g., Böhlke and Bertram (2003) and Böhlke (2004) for FCC polycrystals with isotropic texture. However, the Voigt bound can only be computed for explicit discrete CODF (i.e., superposition of Dirac distributions) and numerical inversion for given viscous exponent n of the nonlinear material laws. Also, the Reuss bound requires for increasing n an increasing number of texture coefficients (up to ∞ for the limit $n \to \infty$) for the present example. The texture dependent expressions of this work might return results beyond the Voigt lower bound of the complementary potential. This is already visible in Fig. 4.5 for $V_1' = 0$, representing the up to fourth-order crystallographic isotropic texture. For isotropic texture, the Voigt bound is well know in literature and delivers for $n \to \infty$ the Taylor factor 3.06 (Taylor, 1938), whereas the present approach exceeds it by around 40% ($\lim_{n\to\infty} \bar{\sigma}_f/\tau_0|_{V_1'=0} \approx 4.29$). For clarity, consider the following two microscopically and macroscopically cubic CODFs

$$
\begin{gathered}
f_k(\boldsymbol{Q}) = \frac{1}{24^2} \sum_{i,j=1}^{24} \delta(\boldsymbol{Q}_i \hat{\boldsymbol{Q}}_k \boldsymbol{Q}_j, \boldsymbol{Q}) \,, \; k \in \{1,2\} \,, \\
\boldsymbol{Q}_i \in S^{\text{cub}} \,, \; \hat{\boldsymbol{Q}}_k = \boldsymbol{Q}\left(\boldsymbol{n} = \frac{1}{\sqrt{3}} \sum_{i=1}^{3} \boldsymbol{b}_i, \omega_k\right) \,, \\
\omega_k \in \left\{\pi, \tfrac{1}{3}\left(2\pi + \cos^{-1}\left(-\tfrac{7}{20}\right)\right)\right\} .
\end{gathered}
\tag{4.29}
$$

The CODF $f_1(\boldsymbol{Q})$ induces $V_1' = -13/27$ (see Lobos et al. (2017)), while $f_2(\boldsymbol{Q})$ induces $V_1' = 0$. The HS, the Voigt and the Reuss bounds for the present example have been computed for the CODFs given in Eq. (4.29). The viscous exponent has been varied as $n = 1, 3, \ldots, 17$. The results are presented in Fig. 4.6. It is clearly visible that the present approach surpasses almost immediately the Voigt bound for the CODFs $f_{1/2}(\boldsymbol{Q})$.

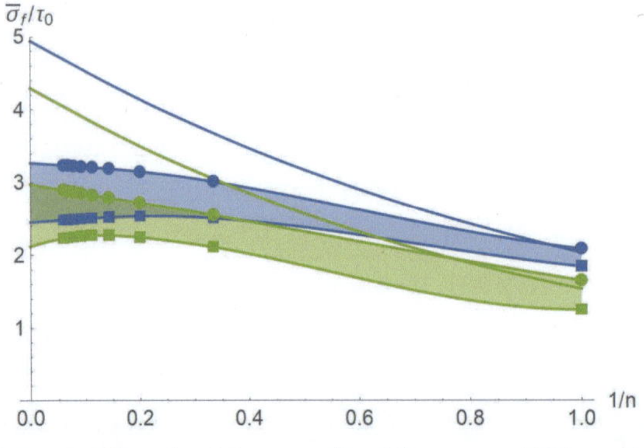

Figure 4.6: Bounds of the normalized flow stress: HS bound depicted by the lines without markers, Voigt bound depicted by the lines with circle markers and Reuss bound depicted by the lines with rectangle markers; the case with the CODF $f_1(\boldsymbol{Q})$ inducing $V_1' = -13/27$ is depicted in blue, while $f_2(\boldsymbol{Q})$ inducing $V_1' = 0$ is depicted in green.

Still, the present approach is a valid bound. The Voigt bound requires numerical inversion to be evaluated in the complementary space, which increases for polycrystals with a large number of orientations. The present approach does not have this disadvantage due to its formulation in complementary space. Additionally the expressions of the present work require only texture coefficients of second- and fourth-order, compared to the Reuss upper bound of the complementary potential. These properties of the results of this work might still be attractive, either for homogenization approaches (to be appropriately computed with higher accuracy by making use of eigenfields in the LCC), or for materials design problems requiring simple bounds in order to *exclude* materials. Again, the reader should keep in mind that the approach of this work has been build for a practical and flexible parametrization of the influence of crystallographic texture. If the texture is given, then the reader should not only consider the present approach, but also the approaches of DeBotton and Ponte Castañeda (1995), Ponte Castañeda (2002), Liu and Ponte Castañeda (2004) and Ponte Castañeda (2015) for further optimized bounds and estimates for known texture.

4.4 Review

In this chapter, applications of the expressions obtained in Chapter 2 and Chapter 3 have been demonstrated.

The first example showed how a polycrystal of a cubic material with orthotropic sample symmetry could be optimized concerning its anisotropic linear elastic properties for a given properties-profile. The materials design approach considered the zeroth-order bounds in the materials selection step. The first- and the second-order bounds were used to delineate the corresponding properties-closure, and the ISC approach was used to search for optimal CODF texture coefficient estimates.

In the second example, the linear thermoelastic properties of a particle reinforced composite were optimized to show how the expressions accounting for eigenstrains can be used. These expressions offer a structure in which anisotropic particles or domains can be incorporated. The anisotropic domains can be, e.g., results from 2-step-homogenization techniques, such that the texture dependent expressions of the present work offer the possibility to design the material orientation of such domains. The thermoelastic composite was assumed to be microscopically and macroscopically transversely isotropic. Its properties were optimized based on the SA1 approximation, which allows for a faster computation in scenarios in which the ISC approach would require more resources.

The last example shows how the results of the present work might be applied in the field of homogenization of nonlinear properties. Based on the LCC method and the dual formulation of the HS bounds, a viscoplastic material with a standard power law has been examined. The LCC method allows obtaining a lower bound with a corresponding polycrystalline linear material. The effective behavior of the polycrystalline linear material is bounded with the texture dependent expressions of the present work, which deliver practical expressions with finite-dimensional parametrization in terms of texture coefficients.

Chapter 5

Summary

The present work approached problems for the materials design of the mechanical properties of textured materials.

For this purpose, in Chapter 2 the description of orientations $\boldsymbol{Q}$ and of the CODF $f(\boldsymbol{Q})$ has been illustrated in terms of Fourier expansions for central distributions (depending on texture eigenvalues λ_α) and for the general case (depending on texture coefficients $\mathbb{V}'_{\langle\alpha\rangle\beta}$). The convex sets of relevant texture eigenvalues (generalizing the results of Lobos et al. (2017)) and coefficients for the orientation average (based on the harmonic decomposition) of a tensorial quantity of r-th-order $\tilde{\mathbb{D}}_{\langle r\rangle}$ have been discussed. The set of texture eigenvalues of the present work offers a trackable option with less complexity than the full set of texture coefficients.

The basic homogenization steps, relevant for the materials of interest in this work, have been sketched in Chapter 3. Since the formal exact non-local effective properties of random materials are impossible to evaluate for general cases, bounds and approximations are searched for. The zeroth-, first- and second-order bounds of linear properties accounting for eigenfields are considered. The zeroth-order bounds of the present work generalize the investigation of Lobos and Böhlke (2016) and offer the possibility to automate the exclusion of materials and selection of material candidates in large databases based on prescribed properties. The texture dependency of the second-order Hashin-Shtrikman bounds for isotropic two-point statistics accounting for eigenfields has been discussed explicitly for arbitrarily anisotropic materials. *Only second- and fourth-order* texture coefficients enter all expressions, extending the results of Böhlke and Lobos (2014) which were valid only for cubic linear elastic materials. These expressions offer *low and finite dimensional* parametrization of all bounds of the present work. Based on the non-diagonal bounds of Lobos and Böhlke (2016), bounds for all eigenfields are obtained. The implications of these results for linear thermoelasticity and nonlinear bounds based on the LCC have been sketched for arbitrarily anisotropic textured materials.

Three examples have been illustrated in Chapter 4 to show possible applications of the texture dependent, low and finite dimensional expressions of this work in the field of materials design. In the first example, a polycrystal of a cubic material with orthotropic texture has been considered. An artificial material database has been used to illustrate the possibilities and difficulties in materials screening based on the zeroth-order bounds of this work. Then, based on the texture dependent results and component bounds of this work, the zeroth-, first- and second-order properties closure of the linear elastic properties have been depicted. These maps help material designers to visualize properties out of range as well as possibly achievable by the actual material. Based on the ISC approach, a potentially optimal texture has been computed, and the maximal possible deviations based on the first- and second-order bounds have been discussed. In the second example, an analogous materials design has been carried out for the linear thermoelastic material properties of a matrix-inclusion composite with microscopic and macroscopic transversally isotropic behavior. This example shows a scenario relevant for fiber reinforced materials. The evaluation of the zeroth-, first- and second-order bounds reflects the limitation of the properties design options. The texture is optimized based on the SA1 approach and maximal possible deviations based on the first- and second-order bounds are computed. The last example illustrates the application of the textured HS bound of this work in the context of nonlinear bounds based on a linear material. The special case of a viscoplastic polycrystal with cubic texture is presented, together with closed-form expressions for the nonlinear material law estimate, as well as the texture dependent bound of the effective flow stress.

The texture dependent results of this work are a contribution to the methods applied in the homogenization and materials design community, which might facilitate solutions to inverse problems of different applications. All expressions have been treated in the context of linear elasticity with eigenfields. However, they might be transferred to other physical problems based on elliptic differential equations.

Future work may focus on the coupling and investigation of the texture dependent expressions of this work for the more recent nonlinear bounds of Ponte Castañeda (2015) with an ansatz for the evolution of texture coefficients, see, e.g., Böhlke (2006). Such a combination of approaches is not only relevant for polycrystals but may also be an interesting approach for viscous nonlinear fiber reinforced composites with evolving orientation distribution of the fibers.

Appendix A

Auxiliary expressions for second- and fourth-order tensors

A.1 Matrix representation of second- and fourth-order tensors

In this work, mainly second- and fourth-order tensors are used for explicit computations. The space of symmetric second-order tensors will be shortly addressed as T_2^s. Every symmetric second-order tensor $\boldsymbol{A} = \sum_{i,j=1}^{3} A_{ij}\boldsymbol{b}_{ij}$ has six degrees of freedom. A harmonic second-order tensor $\boldsymbol{A}$ (i.e., symmetric and fulfilling $\boldsymbol{A} \cdot \boldsymbol{I} = 0$) is parametrized as

$$A_{ij} = \begin{bmatrix} A_{11} & A_{12} & A_{13} \\ A_{12} & A_{22} & A_{23} \\ A_{13} & A_{23} & -A_{11} - A_{22} \end{bmatrix} . \tag{A.1}$$

Based on the following six linearly independent symmetric orthonormal basis tensors

$$\begin{aligned} \boldsymbol{B}_1 &= \boldsymbol{b}_{11} \, , & \boldsymbol{B}_4 &= \tfrac{1}{\sqrt{2}}(\boldsymbol{b}_{23} + \boldsymbol{b}_{32}) \, , \\ \boldsymbol{B}_2 &= \boldsymbol{b}_{22} \, , & \boldsymbol{B}_5 &= \tfrac{1}{\sqrt{2}}(\boldsymbol{b}_{13} + \boldsymbol{b}_{31}) \, , \\ \boldsymbol{B}_3 &= \boldsymbol{b}_{33} \, , & \boldsymbol{B}_6 &= \tfrac{1}{\sqrt{2}}(\boldsymbol{b}_{12} + \boldsymbol{b}_{21}) \, , \end{aligned} \tag{A.2}$$

every $\boldsymbol{A} \in T_2^s$ can be expressed as a linear combination of the chosen basis as

$$\boldsymbol{A} = \sum_{i=1}^{6} A_i \boldsymbol{B}_i \ , \ A_i = \begin{bmatrix} A_{11} & A_{22} & A_{33} & \sqrt{2}A_{23} & \sqrt{2}A_{13} & \sqrt{2}A_{12} \end{bmatrix}^{\mathsf{T}} . \tag{A.3}$$

Every fourth-order tensor $\mathbb{A} = \sum_{i,j,k,l=1}^{3} A_{ijkl} \boldsymbol{b}_{ijkl}$ being left and right symmetric (also referred to as minor symmetric), i.e, $A_{ijkl} = A_{jikl} = A_{ijlk}$, maps from T_2^s to T_2^s and has 36 degrees of freedom. These tensors can be represented as

$$\mathbb{A} = \sum_{i,j=1}^{6} A_{ij} \boldsymbol{B}_i \otimes \boldsymbol{B}_j \tag{A.4}$$

with

$$A_{ij} = \begin{bmatrix} A_{1111} & A_{1122} & A_{1133} & \sqrt{2}A_{1123} & \sqrt{2}A_{1113} & \sqrt{2}A_{1112} \\ A_{2211} & A_{2222} & A_{2233} & \sqrt{2}A_{2223} & \sqrt{2}A_{2213} & \sqrt{2}A_{2212} \\ A_{3311} & A_{3322} & A_{3333} & \sqrt{2}A_{3323} & \sqrt{2}A_{3313} & \sqrt{2}A_{3312} \\ \sqrt{2}A_{2311} & \sqrt{2}A_{2322} & \sqrt{2}A_{2333} & 2A_{2323} & 2A_{2313} & 2A_{2312} \\ \sqrt{2}A_{1311} & \sqrt{2}A_{1322} & \sqrt{2}A_{1333} & 2A_{1323} & 2A_{1313} & 2A_{1312} \\ \sqrt{2}A_{1211} & \sqrt{2}A_{1222} & \sqrt{2}A_{1233} & 2A_{1223} & 2A_{1213} & 2A_{1212} \end{bmatrix} . \tag{A.5}$$

If a minor symmetric tensor $\mathbb{A}$ is additionally major symmetric, i.e., $A_{ijkl} = A_{klij}$, then $\mathbb{A}$ has only 21 degrees of freedom and the matrix Eq. (A.5) is symmetric. A harmonic $\mathbb{A}$ (i.e., completely symmetric and fulfilling $\mathbb{A}[\boldsymbol{I}] = \boldsymbol{0}$) has only 9 degrees of freedom. Such harmonic fourth-order tensors can be parametrized by the symmetric matrix based on

Eq. (A.5) corresponding to a minor and major symmetric fourth-order tensors and the following substitutions

$$
\begin{aligned}
&\{ \\
&A_{1133} \to -A_{1111} - A_{1122} \\
&A_{1212} \to A_{1122} \\
&A_{1213} \to A_{1123} \\
&A_{1233} \to -A_{1112} - A_{1222} \\
&A_{1313} \to -A_{1111} - A_{1122} \\
&A_{1322} \to A_{1223} \\
&A_{1323} \to -A_{1112} - A_{1222} \\
&A_{1333} \to -A_{1113} - A_{1223} \\
&A_{2233} \to -A_{1122} - A_{2222} \\
&A_{2323} \to -A_{1122} - A_{2222} \\
&A_{2333} \to -A_{1123} - A_{2223} \\
&A_{3333} \to A_{1111} + 2A_{1122} + A_{2222} \\
&\}\,.
\end{aligned}
\tag{A.6}
$$

The resulting matrix depends on the components

$$\{A_{1111}, A_{1112}, A_{1113}, A_{1122}, A_{1123}, A_{1222}, A_{1223}, A_{2222}, A_{2223}\} \tag{A.7}$$

which reflect the 9 degrees of freedom of a harmonic $\mathbb{A}$. If the harmonic $\mathbb{A}$ is required to be orthotropic, i.e.,

$$\mathbb{A} = \boldsymbol{Q} \star \mathbb{A} \quad \forall \boldsymbol{Q} \in S^{\mathrm{ort}} \tag{A.8}$$

is imposed based on S^{ort}, defined in Eq. (2.17), then $\mathbb{A}$ has only 3 degrees of freedom and can be represented as $\mathbb{A} = \sum_{i,j=1}^{6} A_{ij} \boldsymbol{B}_i \otimes \boldsymbol{B}_j$ with the replacements

$$\{A_{1112} \to 0, A_{1113} \to 0, A_{1123} \to 0, A_{1222} \to 0, A_{1223} \to 0, A_{2223} \to 0\}\ . \tag{A.9}$$

The orthotropic harmonic $\mathbb{A}$ depends then solely on the three linearly independent components

$$\{A_{1111}, A_{1122}, A_{2222}\} \tag{A.10}$$

which reflect the remaining 3 degrees freedom. The orthotropic harmonic fourth-order tensor can be considered for the parametrization of corresponding texture coefficients, as discussed following Eq. (2.46) and used in Section 4.2.2.

A.2 Cubic fourth-order tensors

The following relations are of practical use for Section 4.2.2. A fourth-order tensor $\mathbb{A}$ is referred to as cubic, if it fulfills

$$\mathbb{A} = \boldsymbol{Q} \star \mathbb{A} \quad \forall \boldsymbol{Q} \in S^{\text{cub}} \tag{A.11}$$

for the cubic symmetry group defined in Eq. (2.17). A minor and major symmetric cubic tensor $\mathbb{A}$ can be expressed into its spectral decomposition, see, e.g., Böhlke and Lobos (2014),

$$\begin{aligned} \mathbb{A} &= \sum_{i=1}^{3} \lambda_i \mathbb{P}_i^{\text{cub}} , \quad \mathbb{P}_1^{\text{cub}} = \mathbb{P}_1 = \frac{1}{2} \boldsymbol{I} \otimes \boldsymbol{I} , \\ \mathbb{D}^{\text{cub}} &= \sum_{i=1}^{3} \boldsymbol{b}_i^{\otimes 4} , \quad \mathbb{P}_2^{\text{cub}} = \mathbb{D}^{\text{cub}} - \mathbb{P}_1^{\text{cub}} , \\ \mathbb{P}_3^{\text{cub}} &= \mathbb{I}^S - (\mathbb{P}_1^{\text{cub}} + \mathbb{P}_2^{\text{cub}}) . \end{aligned} \tag{A.12}$$

It is shortly pointed out, that an isotropic tensors $\mathbb{A} = \lambda_1 \mathbb{P}_1 + \lambda_2 \mathbb{P}_2$ can be expressed as a cubic tensor by $\mathbb{A} = \lambda_1 \mathbb{P}_1^{\text{cub}} + \lambda_2 \mathbb{P}_2^{\text{cub}} + \lambda_2 \mathbb{P}_3^{\text{cub}}$.

The cubic harmonic $\mathbb{A}$ has only one degree of freedom and can be represented as $\mathbb{A} = \sum_{i,j=1}^{6} A_{ij} \boldsymbol{B}_i \otimes \boldsymbol{B}_j$ with

$$A_{ij} = \begin{bmatrix} A_{1111} & -\frac{A_{1111}}{2} & -\frac{A_{1111}}{2} & 0 & 0 & 0 \\ -\frac{A_{1111}}{2} & A_{1111} & -\frac{A_{1111}}{2} & 0 & 0 & 0 \\ -\frac{A_{1111}}{2} & -\frac{A_{1111}}{2} & A_{1111} & 0 & 0 & 0 \\ 0 & 0 & 0 & -A_{1111} & 0 & 0 \\ 0 & 0 & 0 & 0 & -A_{1111} & 0 \\ 0 & 0 & 0 & 0 & 0 & -A_{1111} \end{bmatrix}. \quad \text{(A.13)}$$

The space of cubic harmonic fourth-order tensors is, therefore, one dimensional. A basis tensor $\mathbb{T}'_{\langle 4 \rangle} = \mathbb{T}'$ of this space (used as a reference tensor for the CODF of cubic materials in this work) is given by $\mathbb{T}' = \sum_{i,j=1}^{6} T'_{ij} \boldsymbol{B}_i \otimes \boldsymbol{B}_j$ with

$$T'_{ij} = \sqrt{\frac{2}{15}} \begin{bmatrix} 1 & -\frac{1}{2} & -\frac{1}{2} & 0 & 0 & 0 \\ -\frac{1}{2} & 1 & -\frac{1}{2} & 0 & 0 & 0 \\ -\frac{1}{2} & -\frac{1}{2} & 1 & 0 & 0 & 0 \\ 0 & 0 & 0 & -1 & 0 & 0 \\ 0 & 0 & 0 & 0 & -1 & 0 \\ 0 & 0 & 0 & 0 & 0 & -1 \end{bmatrix}. \quad \text{(A.14)}$$

A.3 Hexagonal second- and fourth-order harmonic tensors

The following relations are of practical use for the discussion concerning Fig. 2.2 and for the example illustrated in Section 4.2.3. A harmonic second-order tensor $\boldsymbol{A}$ is referred to as hexagonal, if it fulfills

$$\boldsymbol{A} = \boldsymbol{Q} \star \boldsymbol{A} \quad \forall \boldsymbol{Q} \in S^{\text{hex}} \quad \text{(A.15)}$$

for the hexagonal symmetry group defined in Eq. (2.17). This implies that the hexagonal harmonic $\boldsymbol{A}$ has only one degree of freedom and can be represented as the linear combination $\boldsymbol{A} = \sum_{i,j=1}^{3} A_{ij} \boldsymbol{b}_{ij}$ with components

$$A_{ij} = \begin{bmatrix} A_{11} & 0 & 0 \\ 0 & A_{11} & 0 \\ 0 & 0 & -2A_{11} \end{bmatrix} \tag{A.16}$$

The space of hexagonal harmonic second-order tensors is, therefore, one dimensional. A basis tensor $\mathbb{T}'_{\langle 2\rangle} = \boldsymbol{T}'$ of this space (used as a reference tensor for the CODF of hexagonal materials in this work) is given by $\boldsymbol{T}' = \sum_{i,j=1}^{3} T'_{ij} \boldsymbol{b}_{ij}$ with

$$T'_{ij} = \frac{1}{\sqrt{6}} \begin{bmatrix} 1 & 0 & 0 \\ 0 & 1 & 0 \\ 0 & 0 & -2 \end{bmatrix}. \tag{A.17}$$

A harmonic fourth-order tensor $\mathbb{A}$ is referred to as hexagonal, if it fulfills

$$\mathbb{A} = \boldsymbol{Q} \star \mathbb{A} \quad \forall \boldsymbol{Q} \in S^{\text{hex}} \tag{A.18}$$

for the hexagonal symmetry group defined in Eq. (2.17). This implies that the hexagonal harmonic $\mathbb{A}$ has only one degree of freedom and can be represented as $\mathbb{A} = \sum_{i,j=1}^{6} A_{ij} \boldsymbol{B}_i \otimes \boldsymbol{B}_j$ with

$$A_{ij} = \begin{bmatrix} A_{1111} & \frac{A_{1111}}{3} & -\frac{4A_{1111}}{3} & 0 & 0 & 0 \\ \frac{A_{1111}}{3} & A_{1111} & -\frac{4A_{1111}}{3} & 0 & 0 & 0 \\ -\frac{4A_{1111}}{3} & -\frac{4A_{1111}}{3} & \frac{8A_{1111}}{3} & 0 & 0 & 0 \\ 0 & 0 & 0 & -\frac{8A_{1111}}{3} & 0 & 0 \\ 0 & 0 & 0 & 0 & -\frac{8A_{1111}}{3} & 0 \\ 0 & 0 & 0 & 0 & 0 & \frac{2A_{1111}}{3} \end{bmatrix}. \tag{A.19}$$

The space of hexagonal harmonic fourth-order tensors is, therefore, one dimensional. A basis tensor $\mathbb{T}'_{\langle 4\rangle} = \mathbb{T}'$ of this space (used as a reference tensor for the CODF of hexagonal materials in this work) is given by $\mathbb{T}' = \sum_{i,j=1}^{6} T'_{ij} \boldsymbol{B}_i \otimes \boldsymbol{B}_j$ with

$$T'_{ij} = \frac{3}{2\sqrt{70}} \begin{bmatrix} 1 & \frac{1}{3} & -\frac{4}{3} & 0 & 0 & 0 \\ \frac{1}{3} & 1 & -\frac{4}{3} & 0 & 0 & 0 \\ -\frac{4}{3} & -\frac{4}{3} & \frac{8}{3} & 0 & 0 & 0 \\ 0 & 0 & 0 & -\frac{8}{3} & 0 & 0 \\ 0 & 0 & 0 & 0 & -\frac{8}{3} & 0 \\ 0 & 0 & 0 & 0 & 0 & \frac{2}{3} \end{bmatrix} . \tag{A.20}$$

Appendix B

Isotropic tensors

B.1 Basic isotropic tensors

In this appendix, the following basic r-th-order isotropic tensor with $r \geq 2$ will be considered

$$\mathbb{B}^{I}_{\langle r \rangle} = \begin{cases} \boldsymbol{I}^{\otimes r/2} & r \text{ even} \\ \boldsymbol{\epsilon} \otimes (\boldsymbol{I}^{\otimes (r-3)/2}) & r \text{ odd} \end{cases}, \tag{B.1}$$

see, e.g., Andrews and Ghoul (1999). Several different tensor transpositions will be used, which will be given as lists, e.g., the list $t = \{1,5,2,6,3,7,4,8\}$ has eight elements, stands for the transposition of a general eight-order tensor $\mathbb{A}_{\langle 8 \rangle}$ and is applied as follows

$$\begin{aligned} \mathbb{A}^{\mathsf{T}_{15263748}}_{\langle 8 \rangle} &= \mathbb{A}^{\mathsf{T}(t)}_{\langle 8 \rangle} \\ &= (A_{i_1 i_2 i_3 i_4 i_5 i_6 i_7 i_8} \boldsymbol{b}_{i_1 i_2 i_3 i_4 i_5 i_6 i_7 i_8})^{\mathsf{T}_{15263748}} \\ &= A_{i_1 i_5 i_2 i_6 i_3 i_7 i_4 i_8} \boldsymbol{b}_{i_1 i_2 i_3 i_4 i_5 i_6 i_7 i_8} \end{aligned} \tag{B.2}$$

The basic isotropic tensor can be used together with the tensor transpositions just introduced in order to generate bases for isotropic tensors of arbitrary order. For a computational approach in *Mathematica*® 11 , see the routines given in Appendix B.4. In order to keep the representation of the isotropic tensor compact, the following notation is introduced. The lists of tensor transpositions $\{t_{ri}\}$, with $i = 1, \dots, d^I_r$, will denote a list of tensor transpositions of the basic isotropic r-th-order tensor $\mathbb{B}^I_{\langle r\rangle}$ which deliver a basis of the d^I_r-dimensional space of r-th-order isotropic tensors. Any isotropic r-th-order tensor $\mathbb{A}^I_{\langle r\rangle}$ can be expressed as a linear combination with coefficients $\{c_i\}$ of the tensor transpositions of $\mathbb{B}^I_{\langle r\rangle}$. For example, the space of isotropic sixth-order tensors is 15-dimensional, i.e., 15 transpositions of $\mathbb{B}^I_{\langle 6\rangle}$

$$\{t_{6i}\} = \{\{1,2,3,4,5,6\},\dots\} \tag{B.3}$$

delivering a basis are needed. An isotropic sixth-order tensor $\mathbb{A}^I_{\langle 6\rangle}$ will be coded as follows

$$\mathbb{A}^I_{\langle 6\rangle} = \sum_{i=1}^{15} c_i(\mathbb{B}^I_{\langle 6\rangle})^{\mathsf{T}(t_{6i})} = \{c_i\} = \{c_1, c_2, \dots, c_{15}\}\,. \tag{B.4}$$

The introduced lists for tensor transpositions and linear combinations of the basic isotropic tensor can be used in the software *Mathematica*® 11 to compute all upcoming tensors, see routines given in Appendix B.4.

The following lists, generated with the routines given in Appendix B.4, deliver bases for the respective tensor orders from fourth-order to eighth-order and will be used for linear combinations of $\mathbb{B}^I_{\langle r\rangle}$ in all upcoming sections. These lists coincide with the tensor transpositions given in Lobos et al. (2017).

Fourth-order: $d^I_4 = 3$.

$$\{t_{4i}\} = \{\{1,2,3,4\},\{1,3,2,4\},\{1,4,2,3\}\} \tag{B.5}$$

Fifth-order: $d^I_5 = 6$.

$$\begin{aligned} &\{t_{5i}\} \\ &= \\ &\{\{1,2,3,4,5\},\{1,2,4,3,5\},\{1,2,5,3,4\},\{1,3,4,2,5\}, \\ &\{1,3,5,2,4\},\{1,4,5,2,3\}\} \end{aligned} \tag{B.6}$$

Sixth-order: $d_6^I = 15$.

$$\begin{aligned}
&\{t_{6i}\} \\
&= \\
&\{\{1,2,3,4,5,6\},\{1,2,3,5,4,6\},\{1,2,3,6,4,5\},\{1,3,2,4,5,6\}, \\
&\{1,3,2,5,4,6\},\{1,3,2,6,4,5\},\{1,4,2,3,5,6\},\{1,4,2,5,3,6\}, \\
&\{1,4,2,6,3,5\},\{1,5,2,3,4,6\},\{1,5,2,4,3,6\},\{1,5,2,6,3,4\}, \\
&\{1,6,2,3,4,5\},\{1,6,2,4,3,5\},\{1,6,2,5,3,4\}\}
\end{aligned} \tag{B.7}$$

Seventh-order: $d_7^I = 36$.

$$\begin{aligned}
&\{t_{7i}\} \\
&= \\
&\{\{1,2,3,4,5,6,7\},\{1,2,3,4,6,5,7\},\{1,2,3,4,7,5,6\}, \\
&\{1,2,4,3,5,6,7\},\{1,2,4,3,6,5,7\},\{1,2,4,3,7,5,6\}, \\
&\{1,2,5,3,4,6,7\},\{1,2,5,3,6,4,7\},\{1,2,5,3,7,4,6\}, \\
&\{1,2,6,3,4,5,7\},\{1,2,6,3,5,4,7\},\{1,2,6,3,7,4,5\}, \\
&\{1,2,7,3,4,5,6\},\{1,2,7,3,5,4,6\},\{1,2,7,3,6,4,5\}, \\
&\{1,3,4,2,5,6,7\},\{1,3,4,2,6,5,7\},\{1,3,4,2,7,5,6\}, \\
&\{1,3,5,2,4,6,7\},\{1,3,5,2,6,4,7\},\{1,3,5,2,7,4,6\}, \\
&\{1,3,6,2,4,5,7\},\{1,3,6,2,5,4,7\},\{1,3,6,2,7,4,5\}, \\
&\{1,3,7,2,4,5,6\},\{1,3,7,2,5,4,6\},\{1,3,7,2,6,4,5\}, \\
&\{1,4,5,2,3,6,7\},\{1,4,5,2,6,3,7\},\{1,4,6,2,3,5,7\}, \\
&\{1,4,6,2,5,3,7\},\{1,4,7,2,3,5,6\},\{1,4,7,2,5,3,6\}, \\
&\{1,5,6,2,3,4,7\},\{1,5,7,2,3,4,6\},\{1,6,7,2,3,4,5\}\}
\end{aligned} \tag{B.8}$$

Eighth-order: $d_8^I = 91$.

$$
\begin{aligned}
&\{t_{8i}\}\\
&=\\
&\{\{1,2,3,4,5,6,7,8\},\{1,2,3,4,5,7,6,8\},\{1,2,3,4,5,8,6,7\},\\
&\{1,2,3,5,4,6,7,8\},\{1,2,3,5,4,7,6,8\},\{1,2,3,5,4,8,6,7\},\\
&\{1,2,3,6,4,5,7,8\},\{1,2,3,6,4,7,5,8\},\{1,2,3,6,4,8,5,7\},\\
&\{1,2,3,7,4,5,6,8\},\{1,2,3,7,4,6,5,8\},\{1,2,3,7,4,8,5,6\},\\
&\{1,2,3,8,4,5,6,7\},\{1,2,3,8,4,6,5,7\},\{1,2,3,8,4,7,5,6\},\\
&\{1,3,2,4,5,6,7,8\},\{1,3,2,4,5,7,6,8\},\{1,3,2,4,5,8,6,7\},\\
&\{1,3,2,5,4,6,7,8\},\{1,3,2,5,4,7,6,8\},\{1,3,2,5,4,8,6,7\},\\
&\{1,3,2,6,4,5,7,8\},\{1,3,2,6,4,7,5,8\},\{1,3,2,6,4,8,5,7\},\\
&\{1,3,2,7,4,5,6,8\},\{1,3,2,7,4,6,5,8\},\{1,3,2,7,4,8,5,6\},\\
&\{1,3,2,8,4,5,6,7\},\{1,3,2,8,4,6,5,7\},\{1,3,2,8,4,7,5,6\},\\
&\{1,4,2,3,5,6,7,8\},\{1,4,2,3,5,7,6,8\},\{1,4,2,3,5,8,6,7\},\\
&\{1,4,2,5,3,6,7,8\},\{1,4,2,5,3,7,6,8\},\{1,4,2,5,3,8,6,7\},\\
&\{1,4,2,6,3,5,7,8\},\{1,4,2,6,3,7,5,8\},\{1,4,2,6,3,8,5,7\},\\
&\{1,4,2,7,3,5,6,8\},\{1,4,2,7,3,6,5,8\},\{1,4,2,7,3,8,5,6\},\\
&\{1,4,2,8,3,5,6,7\},\{1,4,2,8,3,6,5,7\},\{1,4,2,8,3,7,5,6\},\\
&\{1,5,2,3,4,6,7,8\},\{1,5,2,3,4,7,6,8\},\{1,5,2,3,4,8,6,7\},\\
&\{1,5,2,4,3,6,7,8\},\{1,5,2,4,3,7,6,8\},\{1,5,2,4,3,8,6,7\},\\
&\{1,5,2,6,3,4,7,8\},\{1,5,2,6,3,7,4,8\},\{1,5,2,6,3,8,4,7\},\\
&\{1,5,2,7,3,4,6,8\},\{1,5,2,7,3,6,4,8\},\{1,5,2,7,3,8,4,6\},\\
&\{1,5,2,8,3,4,6,7\},\{1,5,2,8,3,6,4,7\},\{1,6,2,3,4,5,7,8\},\\
&\{1,6,2,3,4,7,5,8\},\{1,6,2,3,4,8,5,7\},\{1,6,2,4,3,5,7,8\},\\
&\{1,6,2,4,3,7,5,8\},\{1,6,2,4,3,8,5,7\},\{1,6,2,5,3,4,7,8\},\\
&\{1,6,2,5,3,7,4,8\},\{1,6,2,7,3,4,5,8\},\{1,6,2,7,3,5,4,8\},\\
&\{1,6,2,7,3,8,4,5\},\{1,6,2,8,3,4,5,7\},\{1,7,2,3,4,5,6,8\},\\
&\{1,7,2,3,4,6,5,8\},\{1,7,2,3,4,8,5,6\},\{1,7,2,4,3,5,6,8\},\\
&\{1,7,2,4,3,6,5,8\},\{1,7,2,4,3,8,5,6\},\{1,7,2,5,3,4,6,8\},\\
&\{1,7,2,5,3,6,4,8\},\{1,7,2,6,3,4,5,8\},\{1,7,2,8,3,4,5,6\},\\
&\{1,8,2,3,4,5,6,7\},\{1,8,2,3,4,6,5,7\},\{1,8,2,3,4,7,5,6\},\\
&\{1,8,2,4,3,5,6,7\},\{1,8,2,4,3,6,5,7\},\{1,8,2,4,3,7,5,6\},\\
&\{1,8,2,5,3,4,6,7\},\{1,8,2,5,3,6,4,7\},\{1,8,2,6,3,4,5,7\},\\
&\{1,8,2,7,3,4,5,6\}\}
\end{aligned}
\tag{B.9}
$$

B.2 Isotropic tensors for the harmonic decomposition of a minor symmetric fourth-order tensor

The isotropic tensors $\mathbb{J}_{\langle\alpha\rangle\gamma}$ for the harmonic decomposition Eq. (2.68) are given according to the conventions introduced in Eq. (B.4) as

$$\begin{aligned}
\mathbb{J}_{\langle 5\rangle} &= \{-(3/20),-(3/20),3/10,0,-(3/10),-(3/10)\}\,,\\
\mathbb{J}_{\langle 6\rangle 1} &= \{0,0,1/7,0,0,1/7,0,0,1/7,0,0,0,1/7,1/7,1/7\}\,,\\
\mathbb{J}_{\langle 6\rangle 2} &= \{0,0,-(1/3),0,0,1/6,0,0,1/6,0,0,0,1/6,\\
&\quad 1/6,-(1/3)\}\,,\\
\mathbb{J}_{\langle 6\rangle 3} &= \{0,0,-(1/2),0,0,0,0,0,0,0,0,0,0,0,1/2\}\,,\\
\mathbb{J}_{\langle 7\rangle} &= \{0,0,0,0,0,0,0,0,0,0,0,0,0,0,-(3/4),0,0,0,0,0,\\
&\quad 0,0,0,0,0,0,3/4,0,0,0,0,0,3/4,0,0,0\}
\end{aligned} \tag{B.10}$$

see alternatively Lobos et al. (2017).

B.3 Isotropic tensors $\hat{\mathbb{B}}^I_{\langle 2r\rangle\alpha}$ for $r = 1, 2, 3, 4$

This section presents explicit expressions for the isotropic tensor $\hat{\mathbb{B}}^I_{\langle 2r\rangle\alpha}$ defined in Eq. (2.75). The notation introduced in Appendix B.1 is used in this section for the basic isotropic tensor $\mathbb{B}^I_{\langle 2r\rangle}$, corresponding tensor transpositions and corresponding linear combinations.

Case $r = 1$.

$$\begin{aligned}
\hat{\mathbb{B}}^I_{\langle 2\rangle 0} &= \mathbf{0}\,,\\
\hat{\mathbb{B}}^I_{\langle 2\rangle 1} &= \boldsymbol{I} - \hat{\mathbb{B}}^I_{\langle 2\rangle 0} = \boldsymbol{I}\,.
\end{aligned} \tag{B.11}$$

Case $r = 2$ (see $\{t_{4i}\}$).

$$\begin{aligned}
\hat{\mathbb{B}}^I_{\langle 4\rangle 0} &= \mathbb{I}^I = \{1/3, 0, 0\} = \mathbb{P}_1 \,, \\
\hat{\mathbb{B}}^I_{\langle 4\rangle 1} &= \{0, 1/2, -(1/2)\} = \mathbb{P}_3 \,, \\
\hat{\mathbb{B}}^I_{\langle 4\rangle 2} &= \mathbb{I} - (\hat{\mathbb{B}}^I_{\langle 4\rangle 0} + \hat{\mathbb{B}}^I_{\langle 4\rangle 1}) = \{-(1/3), 1/2, 1/2\} = \mathbb{P}_2 \,.
\end{aligned} \tag{B.12}$$

Case $r = 3$ (see $\{t_{6i}\}$).

$$\begin{aligned}
\hat{\mathbb{B}}^I_{\langle 6\rangle 0} &= \mathbb{I}^I_{\langle 6\rangle} \\
&= \{0, 0, 0, 0, 0, 0, 0, 1/6, -(1/6), 0, -(1/6), 1/6, \\
&\quad 0, 1/6, -(1/6)\} \,, \\
\hat{\mathbb{B}}^I_{\langle 6\rangle 1} &= \{-(1/10), -(1/10), 2/5, -(1/10), 2/5, -(1/10), \\
&\quad 2/5, 0, 0, -(1/10), 0, 0, -(1/10), 0, 0\} \,, \\
\hat{\mathbb{B}}^I_{\langle 6\rangle 2} &= \{1/6, 1/6, -(1/3), 1/6, -(1/3), 1/6, -(1/3), 2/3, \\
&\quad 0, 1/6, 0, -(1/3), 1/6, -(1/3), 0\} \,, \\
\hat{\mathbb{B}}^I_{\langle 6\rangle 3} &= \mathbb{I}_{\langle 6\rangle} - \sum_{\alpha=0}^{2} \hat{\mathbb{B}}^I_{\langle 6\rangle\alpha} \\
&= \{-(1/15), -(1/15), -(1/15), -(1/15), -(1/15), \\
&\quad -(1/15), -(1/15), 1/6, 1/6, -(1/15), 1/6, 1/6, \\
&\quad -(1/15), 1/6, 1/6\} \,.
\end{aligned} \tag{B.13}$$

Case $r = 4$ (see $\{t_{8i}\}$). The tensor $\hat{\mathbb{B}}^I_{\langle 8\rangle 0}$ matches exactly the identity on isotropic fourth-order tensors $\mathbb{I}^I_{\langle 8\rangle} = \sum_{i=1}^{3} (\mathbb{P}_i/\|\mathbb{P}_i\|)^{\otimes 2}$, also given in components in Morawiec (1989; 1994).

$$\begin{aligned}
&\hat{\mathbb{B}}^I_{\langle 8\rangle 0} = \mathbb{I}^I_{\langle 8\rangle} \\
&= \\
&\{2/15, -(1/30), -(1/30), 0, 0, 0, 0, 0, 0, 0, 0, 0, 0, 0, 0, \\
&-(1/30), 2/15, -(1/30), 0, 0, 0, 0, 0, 0, 0, 0, 0, 0, 0, 0, \\
&-(1/30), -(1/30), 2/15, 0, 0, 0, 0, 0, 0, 0, 0, 0, 0, 0, 0, \\
&0, 0, 0, 0, 0, 0, 0, 0, 0, 0, 0, 0, 0, 0, 0, 0, 0, 0, 0, 0, 0, \\
&0, 0, 0, 0, 0, 0, 0, 0, 0, 0, 0, 0, 0, 0, 0, 0, 0, 0, 0, 0, 0\} \,,
\end{aligned} \tag{B.14}$$

$$
\begin{aligned}
&\hat{\mathbb{B}}^{I}_{\langle 8\rangle 1}\\
&=\\
&\{0,0,0,0,1/10,-(1/10),0,1/10,-(1/10),-(1/10),-(1/10),\\
&3/10,1/10,1/10,-(3/10),0,0,0,1/10,0,-(1/10),-(1/10),\\
&-(1/10),3/10,0,1/10,-(1/10),1/10,-(3/10),1/10,0,0,0,\\
&1/10,-(1/10),0,-(1/10),3/10,-(1/10),1/10,-(3/10),1/10,\\
&0,1/10,-(1/10),-(1/10),-(1/10),3/10,-(1/10),3/10,\\
&-(1/10),3/10,0,0,-(1/10),0,0,-(1/10),0,1/10,0,-(1/10),\\
&1/10,-(1/10),0,-(3/10),0,1/10,0,0,1/10,1/10,0,-(1/10),\\
&-(3/10),1/10,1/10,1/10,0,-(1/10),0,-(3/10),1/10,1/10,\\
&1/10,0,-(1/10),1/10,0,-(1/10),0\}\ ,
\end{aligned}
\tag{B.15}
$$

$$
\begin{aligned}
&\hat{\mathbb{B}}^{I}_{\langle 8\rangle 2}\\
&=\\
&\{-(10/21),4/21,4/21,2/21,-(1/14),-(1/14),2/21,-(1/14),\\
&-(1/14),-(1/14),-(1/14),11/42,-(1/14),-(1/14),11/42,\\
&4/21,-(10/21),4/21,-(1/14),2/21,-(1/14),-(1/14),\\
&-(1/14),11/42,2/21,-(1/14),-(1/14),-(1/14),11/42,\\
&-(1/14),4/21,4/21,-(10/21),-(1/14),-(1/14),2/21,\\
&-(1/14),11/42,-(1/14),-(1/14),11/42,-(1/14),2/21,\\
&-(1/14),-(1/14),-(1/14),-(1/14),11/42,-(1/14),11/42,\\
&-(1/14),11/42,0,0,-(1/14),0,0,-(1/14),0,-(1/14),2/21,\\
&-(1/14),-(1/14),-(1/14),2/21,11/42,0,-(1/14),0,0,\\
&-(1/14),-(1/14),2/21,-(1/14),11/42,-(1/14),-(1/14),\\
&-(1/14),0,-(1/14),2/21,11/42,-(1/14),-(1/14),-(1/14),\\
&2/21,-(1/14),-(1/14),0,-(1/14),2/21\}\ ,
\end{aligned}
\tag{B.16}
$$

$$\hat{\mathbb{B}}^I_{\langle 8\rangle 3} = \{1/2, 0, -(1/2), -(1/6), -(1/10), 7/20, -(1/6), 2/5, -(3/20), 1/10, 1/10, -(23/60), 3/20, 3/20, -(17/60), 0, 0, 0, -(1/10), 1/12, 1/10, 1/10, -(3/20), -(2/15), 1/12, 3/20, -(3/20), -(1/10), -(1/30), 3/20, -(1/2), 0, 1/2, 2/5, -(3/20), -(1/6), 1/10, -(2/15), -(3/20), 3/20, -(8/15), 2/5, -(1/6), 3/20, 1/10, 7/20, 1/10, -(19/30), -(3/20), -(2/15), 1/10, -(23/60), 1/2, 1/4, -(3/20), 1/2, -(1/4), 7/20, -(1/4), 3/20, -(1/6), 1/10, 3/20, -(3/20), 1/12, -(17/60), 1/4, 2/5, -(1/4), -(1/4), -(1/10), -(1/10), -(1/6), 7/20, -(1/30), 3/20, -(1/10), 3/20, -(1/4), 1/10, -(1/6), -(1/30), -(1/10), 3/20, -(1/10), 1/12, 1/10, 3/20, -(1/4), 1/10, -(1/6)\}\,, \tag{B.17}$$

$$\hat{\mathbb{B}}^I_{\langle 8\rangle 4} = \mathbb{I}_{\langle 8\rangle} - \sum_{\alpha=0}^{3} \hat{\mathbb{B}}^I_{\langle 8\rangle\alpha} = \{-(11/70), -(11/70), 12/35, 1/14, 1/14, -(5/28), 1/14, -(3/7), 9/28, 1/14, 1/14, -(5/28), -(5/28), -(5/28), 9/28, -(11/70), 12/35, -(11/70), 1/14, -(5/28), 1/14, 1/14, 9/28, -(3/7), -(5/28), -(5/28), 9/28, 1/14, 1/14, -(5/28), 12/35, -(11/70), -(11/70), -(3/7), 9/28, 1/14, 1/14, -(3/7), 9/28, -(5/28), 4/7, -(3/7), 1/14, -(5/28), 1/14, -(5/28), 1/14, 1/14, 9/28, -(3/7), 1/14, -(5/28), 1/2, -(1/4), 9/28, -(1/2), 1/4, -(5/28), 1/4, -(5/28), 1/14, 1/14, -(5/28), 9/28, -(5/28), 9/28, -(1/4), -(3/7), 1/4, 1/4, 1/14, 1/14, 1/14, -(5/28), 1/14, -(5/28), 1/14, -(5/28), 1/4, 1/14, 1/14, 1/14, 1/14, -(5/28), 1/14, -(5/28), 1/14, -(5/28), 1/4, 1/14, 1/14\}\,. \tag{B.18}$$

B.4 Routines for *Mathematica*® 11

The following routines for *Mathematica*® 11 can be used in order to generate a basis for isotropic tensors and compute coefficients of a given isotropic tensor in respect to the generated bases.

The author recommends the readers copying and pasting the upcoming lines of code to double check the pasted lines. Some PDF readers might have problems copying some symbols and might copy `{i,r-1}` as `{i,r1}`.

Table B.1: Auxiliary routines 1

```
(*Kronecker delta kd and permutation symbol ps*)
kd = IdentityMatrix[3, SparseArray];
ps = LeviCivitaTensor[3];
(*Dyadic power*)
dpow[A_, r_] := Block[{temp}, temp = A;
    Do[temp = TensorProduct[temp, A], {i, r - 1}];
    temp];
dpow::usage = "dpow[A,r] computes the dyadic
    power of the tensor A, i.e., A
    \[TensorProduct]...\[TensorProduct]A (r times).";
(*Generation of isotropic basic tensor for even
    and odd order r*)
Biso[r_?EvenQ] := dpow[kd, r/2];
Biso[r_?OddQ] := TensorProduct[ps,
    dpow[kd, (r - 3)/2]];
Biso::usage = "Biso[r] computes the basic
    r-th-order isotropic tensor.";
```

For the Example 1 given in Table B.3, the auxiliary and main routines given in Table B.1 and Table B.2 are required. In Example 1, the lists with all tensor transpositions up to eighth-order given in Appendix B.1 are generated and saved in the symbol `isoTTs`. The lists for, e.g., eight-order can be extracted with `isoTTs[[8]]`, see Eq. (B.9).

Table B.2: Main routine 1

```
(*Generate set of basis tensor transpositions for
    isotropic r-th-order tensors*)
isoTT[r_] := Module[{p, m1, m2},
    p = Permutations@Range@r;
    m1 = Table[Flatten@TensorTranspose[Biso[r],
         p[[i]]], {i, Length@p}];
    m2 = DeleteDuplicates@m1;
    p = p[[Flatten[Position[m1, #, 1, 1] & /@ m2]]];
    p = p[[
        Flatten[Position[#,
        Except[0, _?NumericQ], 1, 1] &
        /@ RowReduce@Transpose@m2]]]];
isoTT::usage = "isoTT[r] computes a set of tensor
    transpositions of the basic isotropic r-th-order
    tensor Biso[r]. The computed set of tensor
    transpositions deliver together with Biso[r]
    a basis of the set of isotropic r-th-order
    tensors.";
```

Table B.3: Example 1

```
(*Generate transpositions up to eight-order*)
isoTTs = Table[isoTT@i, {i, 8}];
```

For the Example 2 given in Table B.6, the auxiliary and main routines given in Table B.4 and Table B.5 are required. In Example 2, the coefficients for the representation of $\hat{\mathbb{B}}^{I}_{\langle 2r\rangle\alpha}$ for $r = 3$ and $\alpha = 2$ are determined. The result of this computation is the corresponding list given in Eq. (B.13).

Table B.4: Auxiliary routines 2

```
(*Dirichlet kernels and related functions*)
Dk[a_] := 1 + 2*Sum[Cos[k*om], {k, a}];
muh[a_] := Dk[a]/(1 + 2*a);
(*Dummy rotation Q in terms of rotation angle
    om and rotation axis nv*)
nv = Array[nc, {3}];
Q = (Cos[om]*kd - Sin[om]*ps.nv
    + (1 - Cos[om])*dpow[nv, 2]);
(*Integration over SO(3) for rotation angle
    om and rotation axis nv*)
s = Sin[om/2]^2/(2 Pi^2);
intoms2[f_] := Integrate[Integrate[f,
    Element[nv, Sphere[3]]]*s, {om, 0, Pi}];
intoms2::usage = "intoms2[f] computes the
    integral over SO(3) for a quantity f
    depending on the orientation angle om
    and the rotation axis nv.";
(*Rayleigh power*)
rpow[A_, r_] := TensorTranspose[dpow[A, r]
    , Flatten@Table[{i, r + i}, {i, r}]];
rpow::usage = "rpow[A,r] computes the
    Rayleigh power of the tensor A to the r.";
```

Table B.5: Main routine 2

```
(*Compute integral over SO(3) for Bisoh*)
Bisoh[r_, a_] := intoms2[
   Flatten[rpow[Q, r]]*(1 + 2*a)^2*muh[a]];
Bisoh::usage = "Bisoh[r,a] computes
   the isotropic 2r-th-order tensor for given
   r based on muh[a].";
(*Compute coefficients for linear combinations
   for on transpositions isoTTs for fixed r
   and given isotropic tensor rhs*)
lcBiso[r_, rhs_] := Module[
   {b},
   b = Transpose@Table[Flatten
      @TensorTranspose[Biso[2*r],
      isoTTs[[2 r, i]]]
      ,{i,Length@isoTTs[[2 r]]}];
   LinearSolve[b, rhs]];
lcBiso::usage = "lcBiso[r,rhs] computes the
   coefficients for the linear combination
   in order to represent the given isotropic
   tensor rhs in terms of the transpositions
   isoTTs of the basic 2r-th-order isotropic
   tensor.";
```

Table B.6: Example 2

```
{r, a} = {3, 2};
lcBiso[r, Bisoh[r, a]]
```

Appendix C

Quadratic polynomials

C.1 Local notation

In this appendix, a compact vector matrix notation will be used for simplicity. The space of real valued numbers is denoted as R. Italic symbols, e.g., c, γ and y, will denote scalars in R. Underlined symbols, e.g., $\underline{x}, \underline{b}$ and $\underline{\beta}$, will denote vectors in R^n. Double underlined symbols, e.g., $\underline{\underline{V}}, \underline{\underline{D}}$ and $\underline{\underline{M}}$, will denote quadratic matrices in in $R^{n\times n}$. The symbol $\underline{\underline{A}}$ will be used for symmetric matrices in $R^{n\times n}$. The scalar product between vectors, e.g., $\underline{x}$ and $\underline{b}$ will be simply noted as $\underline{x}^\mathsf{T}\underline{b}$.

C.2 Homogeneous quadratic polynomials

Positive semidefinite quadratic polynomials. We consider in this section homogeneous quadratic polynomials q

$$q = \underline{x}^\mathsf{T}\underline{\underline{A}}\,\underline{x} = \sum_{i,j=1}^{n} x_i A_{ij} x_j \tag{C.1}$$

with symmetric matrix $\underline{\underline{A}}$. The polynomial q is non-negative, i.e., $q \geq 0\ \forall \underline{x}$, if and only if $\underline{\underline{A}}$ is positive semidefinite (PSD), shortly denoted as $0 \leq \underline{\underline{A}}$. A matrix $\underline{\underline{A}}$ is positive semidefinite if and only if all principal minors are non-negative.

For $q^+ = \underline{x}^\mathsf{T} \underline{\underline{A}}^+ \underline{x}$, $q \leq q^+ \; \forall \underline{x}$ is fulfilled if and only if $\underline{\underline{A}}^+ - \underline{\underline{A}}$ is PSD, shortly denoted as $\underline{\underline{A}} \leq \underline{\underline{A}}^+$.

Bounds of matrix components. The following matrix component bounds are relevant for the bounding of material properties in case of given upper and lower bounds, described in Section 3.4. For $0 \leq \underline{\underline{A}} \leq \underline{\underline{A}}^+$, it possible to bound the coefficients of unknown $\underline{\underline{A}}$ for given $\underline{\underline{A}}^+$. This is achieved by consideration of simple necessary conditions. The simplest necessary conditions for $0 \leq \underline{\underline{A}} \leq \underline{\underline{A}}^+$ are obtained with vectors $x_i \neq 0$ and $x_j = 0 \forall j \neq i$. This delivers the necessary conditions for the main diagonal components of $\underline{\underline{A}}$

$$0 \leq A_{ii} \leq A_{ii}^+ \qquad \forall i = 1, \ldots, n \,. \tag{C.2}$$

The off-diagonal components of $\underline{\underline{A}}$ can be bounded as in Lobos and Böhlke (2016) by consideration of the vectors $x_i \neq 0, x_j \neq 0, x_k = 0 \; \forall k \neq i, j$. This delivers the necessary conditions

$$0 \leq \begin{bmatrix} A_{ii} & A_{ij} \\ A_{ij} & A_{jj} \end{bmatrix} \leq \begin{bmatrix} A_{ii}^+ & A_{ij}^+ \\ A_{ij}^+ & A_{jj}^+ \end{bmatrix} . \tag{C.3}$$

For 2×2 matrices, the maximum and minimum of A_{ij} can be obtained analytically, see Lobos and Böhlke (2016), which delivers the off-diagonal bounds

$$\begin{gathered} B_{ij}^- \leq A_{ij} \leq B_{ij}^+ \,, \quad B_{ij}^\pm = \mu_{ij} \pm \Delta_{ij} \,, \\ \mu_{ij} = \tfrac{1}{2} A_{ij}^+ \,, \quad \Delta_{ij} = \tfrac{1}{2} \sqrt{A_{ii}^+ A_{jj}^+} \,. \end{gathered} \tag{C.4}$$

For cases with $\underline{\underline{A}}^- \leq \underline{\underline{A}} \leq \underline{\underline{A}}^+$ and given $\underline{\underline{A}}^\pm$, the problem can be reformulated as $0 \leq (\underline{\underline{A}} - \underline{\underline{A}}^-) \leq (\underline{\underline{A}}^+ - \underline{\underline{A}}^-)$, solved with Eq. (C.4) and given again explicitly in terms of the components of the considered matrix $\underline{\underline{A}}$ as follows

$$
\begin{aligned}
B_{ij}^- \le A_{ij} \le B_{ij}^+ \,, \quad & B_{ij}^{\pm} = \mu_{ij} \pm \Delta_{ij} \,, \\
\mu_{ij} = \tfrac{1}{2}(A_{ij}^+ + A_{ij}^-) \,, \quad & \Delta_{ij} = \tfrac{1}{2}\sqrt{(A_{ii}^+ - A_{ii}^-)(A_{jj}^+ - A_{jj}^-)} \,.
\end{aligned}
\tag{C.5}
$$

Upper bound problem. The following problem is relevant for the zeroth-order bonds discussed in Section 3.4.2. Consider the problem of finding a symmetric matrix $\underline{\underline{B}}$ for given matrices $\{\underline{\underline{A}}_i\} = \{\underline{\underline{A}}_1, \underline{\underline{A}}_2, \dots, \underline{\underline{A}}_n\}$, $\underline{\underline{A}}_i \in R^{n\times n}$ fulfilling $\underline{x}^{\mathsf{T}} \underline{\underline{A}}_i \, \underline{x} \le \underline{x}^{\mathsf{T}} \underline{\underline{B}} \, \underline{x} \; \forall \underline{x} \forall i$. The set of possible upper bounds for the given matrices

$$
\mathcal{B} = \{\underline{\underline{B}} \mid \underline{x}^{\mathsf{T}} \underline{\underline{A}}_i \, \underline{x} \le \underline{x}^{\mathsf{T}} \underline{\underline{B}} \, \underline{x} \; \forall \underline{x} \forall i\} \tag{C.6}
$$

is a convex set since any convex combination of $\underline{\underline{B}}_1 \in \mathcal{B}$ and $\underline{\underline{B}}_2 \in \mathcal{B}$ is also an element of $\mathcal{B}$. A unique / optimal upper bound does not exist unless some order for elements of $\mathcal{B}$ is imposed. For two matrices $\underline{\underline{C}}$ and $\underline{\underline{D}}$, $\underline{\underline{C}} \le \underline{\underline{D}}$ (i.e., $\underline{\underline{D}} - \underline{\underline{C}}$ is PSD) implies $\gamma_i \le \delta_i \; i = 1, \dots, n$ for the eigenvalues γ_i of $\underline{\underline{C}}$ and δ_i of $\underline{\underline{D}}$ in increasing order, see Horn and Johnson (1990). From these necessary conditions, further necessary conditions for $\underline{\underline{C}} \le \underline{\underline{D}}$ are immediately derivable, e.g., $\mathrm{tr}\left(\underline{\underline{C}}\right) \le \mathrm{tr}\left(\underline{\underline{D}}\right)$, $\det\left(\underline{\underline{C}}\right) \le \det\left(\underline{\underline{D}}\right)$ and many others. This motivates the simplest optimal upper bound $\underline{\underline{B}}^+$ as the upper bound with trace being smaller than all upper bounds, i.e., $\underline{\underline{B}}^+$ is defined as

$$
\underline{\underline{B}}^+ = \arg \min_{\underline{\underline{B}} \in \mathcal{B}} \mathrm{tr}\left(\underline{\underline{B}}\right) \,. \tag{C.7}
$$

It should be remarked that the choice for the trace is a practical choice, as used without further discussion in Nadeau and Ferrari (2001) in the context of zeroth-order bounds of linear elastic material behavior. But, alternative optimal bounds can be defined on other invariants, e.g., the determinant, see Lobos and Böhlke (2016) for a discussion in the context of zeroth-order bounds of linear elastic material behavior. It should also be noted that the solution of the minimization of the trace and the solution of the minimization of the determinant differ, in general. In this work, we choose the trace in order to formulate a simpler optimization problem, since the trace is a linear function.

The optimization problem Eq. (C.7) belongs to the positive semidefinite programming class. The objective function in this optimization problem is linear / strictly increasing in $\underline{\underline{B}}$, and the underlying set of variables is convex, such that numerical interior point algorithms searching for a local minimum suffice. Alternatively, this positive semidefinite programming problem can be treated with slacked variables, see Lourenço et al. (2016).

As an example, consider the following matrices

$$\underline{\underline{A}}_1 = \begin{bmatrix} 117 & 42 \\ 42 & 17 \end{bmatrix}, \quad \underline{\underline{A}}_2 = \begin{bmatrix} 26 & 48 \\ 48 & 104 \end{bmatrix} \tag{C.8}$$

For illustrative reasons, we will consider diagonal matrices

$$\underline{\underline{B}} = \begin{bmatrix} b_1 & 0 \\ 0 & b_2 \end{bmatrix} \tag{C.9}$$

for this first example. The set $\mathcal{B}$ is described now based on the non-negative principal minors of the difference matrices $\underline{\underline{B}} - \underline{\underline{A}}_i$

$$\begin{aligned} \mathcal{B}_1 &= \{\underline{\underline{B}} \mid b_2 \geq 17 \wedge b_1 (b_2 - 17) + 225 \geq 117 b_2\}, \\ \mathcal{B}_2 &= \{\underline{\underline{B}} \mid b_2 \geq 104 \wedge b_1 (b_2 - 104) + 400 \geq 26 b_2\}, \\ \mathcal{B} &= \mathcal{B}_1 \cap \mathcal{B}_2 . \end{aligned} \tag{C.10}$$

The regions $\mathcal{B}_1$, $\mathcal{B}_2$ and $\mathcal{B}$ are depicted in Fig. C.2.1 in orange, blue and green respectively while a contour map of $\operatorname{tr}(\underline{\underline{B}})$ is given in the background in gray colors (lighter colors indicate higher values of $\operatorname{tr}(\underline{\underline{B}})$).

The matrices $\underline{\underline{B}}$ with respective minimum traces in $\mathcal{B}_1$, $\mathcal{B}_2$ and $\mathcal{B}$ are depicted by the red, blue and green points, respectively. This explicit example has been chosen in order to motivate two points. The first one is that the solutions of the separate optimization problem differ, i.e., the solution in $\mathcal{B}$ cannot be obtained from the respective solutions in the $\mathcal{B}_i$, in general. And the second point is that for the special case of matrices $\underline{\underline{B}}$ with fixed eigenvectors, as in the current example, taking the maximum eigenvalues of the solutions of each $\mathcal{B}_i$ in each eigenspace suffices in order to obtain an element of $\mathcal{B}$ (e.g., the purple point in Fig. C.2.1). But this upper bound will not have minimum trace in $\mathcal{B}$, in general. In higher dimensional spaces, this optimization problem shows several complications due to the increasing number of constraints.

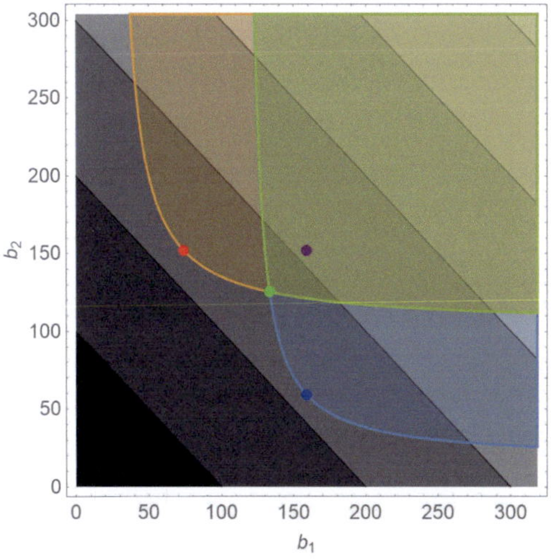

Figure C.2.1: Regions $\mathcal{B}_1$ (orange), $\mathcal{B}_2$ (blue) and $\mathcal{B}$ (green) of the 2D example; a contour map of the trace is given in the backgroud in gray colors, where lighter colors indicate higher trace values; the elements with minimum trace in $\mathcal{B}_1$, $\mathcal{B}_2$ and $\mathcal{B}$ are marked with the red, blue and green points, respectively; the purple points marks the point by taking the maximum values in respective common eigenspaces of the separate solutions in $\mathcal{B}_1$ and $\mathcal{B}_2$.

In Table C.1 three implementations of algorithms in *Mathematica*® 11 for the computation on an upper bound symmetric matrix B for given list of matrices As are given. In Table C.2, the discussed 2D example of Fig. C.2.1 is computed with its output lines given in Table C.3.

Table C.1: Routines for computation of upper bound

```
PrincipalMinors[A_]:=Diagonal@Minors[A,#]&/@Range@Length@A;
PSDCond[A_]:=#>=0&/@Flatten@PrincipalMinors@A;
UpperBoundTr[As_,B_]:=Module[
  {cond,f,vars,min},
  cond=And@@PSDCond[B-#]&/@As;
  f=Tr@B;
  vars=Variables@B;
  min=Minimize[{f,cond},vars];
  FullSimplify@min[[2]]
];
NUpperBoundTr[As_,B_]:=Module[
  {cond,f,vars,min},
  cond=And@@PSDCond[B-#]&/@As;
  f=Tr@B;
  vars=Variables@B;
  min=FindMinimum[{f,cond},vars,Method->"InteriorPoint"];
  min[[2]]
];
NUpperBoundTrSlack[As_,B_]:=Module[
  {cond,f,vars,min,Ss,s},
  Ss=UpperTriangularize/@Array[s,Dimensions@As];
  cond=DeleteCases[Flatten[UpperTriangularize/@
    (ConstantArray[B,Length@As]-As-(Transpose[#].#&/@Ss))],0];
  cond=Thread[cond==0];
  f=Tr@B;
  vars=Join[Variables@B,Variables@Ss];
  vars={#,1.}&/@vars;
  min=FindMinimum[{f,cond},vars];
  First@Solve[B==(B/.min[[2]])]
];
MinEV[As_,B_]:=Min[Eigenvalues@(B-#)&/@As];
UpperBoundQ[As_,B_]:=0<=Min[Eigenvalues[B-#]&/@As];
```

Table C.2: Upper bound example

```
As = {{{117, 42}, {42, 17}}, {{26, 48}, {48, 104}}};
B = DiagonalMatrix@{b1, b2};
bs = Through[{
  UpperBoundTr,
  NUpperBoundTr,
  NUpperBoundTrSlack}[As, B]]
Bs = B /. bs;
UpperBoundQ[As, #] & /@ Bs
MinEV[As, #] & /@ Bs
```

Table C.3: Output lines for upper bound example

```
Out[1] = {
  {b1->1/58 (3967+(14153689)^(1/2))
   ,b2->1/182 (11551+3(14153689)^(1/2))}
  ,{b1->133.261,b2->125.48}
  ,{b1->133.261,b2->125.48}}
Out[2] = {True, True, False}
Out[3] = {0,4.84361*10^(-8),-5.25851*10^(-9)}
```

The symbol bs in Table C.2 carries out all computation routines. The routine UpperBoundTr computes the analytic solution, which for the 2D example is given in the first entry of the list of Out[1] of Table C.3. The second and third entry of Out[1] are the results of the numeric routines NUpperBoundTr (based on conditions for the principal minors and an interior point algorithm) and NUpperBoundTrSlack (based on slack variables for positive semidefinite programming). All solutions stored in bs are then stored in the symbol Bs (creating a list with three matrices, the solutions of the respective routines). These three solutions are then tested with the routine UpperBoundQ which returns in Out[2] in Table C.3 two true statements and a false one for the solution of NUpperBoundTrSlack. This false statement is due to the minimum negative eigenvalue -5.26×10^{-9} of one of the difference matrices $\underline{\underline{B}} - \underline{\underline{A}}_i$ for `NUpperBoundTrSlack`, which might be consider negligible in a numeric sense.

In the author's experience, for higher dimensional problems, routines based on the slack variable formulation, as in the presented routine `NUpperBoundTrSlack`, seems to work faster and more accurately than routines similar to presented `NUpperBoundTr`, which has problems finding feasible points for increasing number of constraints. Problem reformulations based on the Gershgorin circle theorem and diagonal dominance of matrices are possible, but, in the author's experience, slower and less accurate than the results obtained with the slack variable formulation.

C.3 Quadratic polynomials

The following relations are needed for the bounds of linear properties with eigenfields, see Eq. (3.52). We consider quadratic polynomials p

$$p = \underline{x}^{\mathsf{T}} \underline{\underline{A}}\, \underline{x} + 2\underline{x}^{\mathsf{T}} \underline{b} + c\,. \tag{C.11}$$

For p in order to be non-negative, i.e., $p \geq 0\ \forall \underline{x}$, conditions on $\underline{\underline{A}}, \underline{b}$ and c have to be fulfilled. Every symmetric $\underline{\underline{A}}$ can be diagonalized as $\underline{\underline{A}} = \underline{\underline{V}}^{\mathsf{T}} \underline{\underline{D}}\, \underline{\underline{V}}$, with diagonal $\underline{\underline{D}}$ with $\lambda_i, i = 1, \ldots, n$ being the eigenvalues $\underline{\underline{A}}$. Further, we assume that the last m eigenvalues vanish. The polynomial p can be expressed with $\underline{\xi} = \underline{\underline{V}}\, \underline{x}$ and $\underline{\beta} = \underline{\underline{V}}^{\mathsf{T}}\, \underline{b}$ as

$$\begin{aligned} p &= \underline{\xi}^{\mathsf{T}} \underline{\underline{D}}\, \underline{\xi} + 2\underline{\xi}^{\mathsf{T}} \underline{\beta} + c \\ &= \sum_{i=1}^{n-m} (\lambda_i \xi_i^2 + 2\xi_i\beta_i) + c + \sum_{i=n-m+1}^{n} 2\xi_i\beta_i \\ &= \sum_{i=1}^{n-m} \left(\lambda_i \left[\xi_i + \frac{\beta_i}{\lambda_i} \right]^2 \right) + c - \sum_{i=1}^{n-m} \frac{\beta_i^2}{\lambda_i} + \sum_{i=n-m+1}^{n} 2\xi_i\beta_i\,. \end{aligned} \tag{C.12}$$

It is now visible, that the condition for non-negativity $p \geq 0 \forall \underline{x}$ is fulfilled if and only if

$$\begin{aligned} & \lambda_i \geq 0 \quad i = 1, \ldots, n-m \\ \wedge \quad & \beta_i = 0 \quad i = n-m+1, \ldots, n \\ \wedge \quad & \sum_{i=1}^{n-m} \frac{\beta_i^2}{\lambda_i} \leq c \,, \end{aligned} \tag{C.13}$$

such that $\underline{\underline{A}}$ has to be PSD, c has to be non-negative and the components of $\underline{b}$ have to fulfill the constrains in Eq. (C.13) generated by $\underline{\underline{A}}$ and c. For $c = 0$, it follows immediately that $\underline{b} = \underline{0}$ has to be fulfilled for $p \geq 0 \forall \underline{x}$ to hold.

Now, p is assumed non-negative, i.e., $\underline{\underline{A}}, \underline{b}$ and c fulfilled the derived conditions. The symmetric block matrix $\underline{\underline{M}} \in R^{(n+1)\times(n+1)}$

$$\underline{\underline{M}} = \begin{bmatrix} \underline{\underline{A}} & \underline{b} \\ \underline{b}^{\mathsf{T}} & c \end{bmatrix} \tag{C.14}$$

is now investigated. For the vector $\underline{z} = [\underline{x}, y]^{\mathsf{T}}$ with real valued scalar y, the quadratic form of $\underline{\underline{M}}$ is considered

$$\underline{z}^{\mathsf{T}} \underline{\underline{M}}\, \underline{z} = \underline{x}^T \underline{\underline{A}}\, \underline{x} + 2\underline{x}^{\mathsf{T}}(y\underline{b}) + cy^2 \,. \tag{C.15}$$

This quadratic form is non-negative for all $\underline{x}$ and y, i.e., for all $\underline{z}$. This is directly proven by consideration of the conditions Eq. (C.13) for $(y\underline{b})$ and cy^2 which are fulfilled for all y, since y enters linearly the conditions Eq. $(C.13)_2$ and as y^2 both sides of the inequalities in Eq. $(C.13)_3$. Therefore, $\underline{\underline{A}}, \underline{b}$ and c fulfilling Eq. (C.13) imply PSD $\underline{\underline{M}}$. This statement can be given compactly by defining $\underline{z}_1 = [\underline{x}, 1]^{\mathsf{T}}$ and considering $\underline{z}_1^{\mathsf{T}} \underline{\underline{M}}\, \underline{z}_1 = p$ as

$$0 \leq \underline{z}_1^{\mathsf{T}} \underline{\underline{M}}\, \underline{z}_1 \, \forall \underline{z}_1 \Rightarrow 0 \leq \underline{z}^{\mathsf{T}} \underline{\underline{M}}\, \underline{z} \, \forall \underline{z} \Leftrightarrow 0 \leq \underline{\underline{M}} \,. \tag{C.16}$$

For given $p^{\pm}$ fulfilling $p^{-} \leq p \leq p^{+}$, corresponding symmetric block matrices $\underline{\underline{M}}$ and $\underline{\underline{M}}^{\pm}$ can be constructed. For these matrices

$$\underline{z}_1^{\mathsf{T}} \underline{\underline{M}}^{-} \underline{z}_1 \leq \underline{z}_1^{\mathsf{T}} \underline{\underline{M}} \underline{z}_1 \leq \underline{z}_1^{\mathsf{T}} \underline{\underline{M}}^{+} \underline{z}_1 \quad \forall \underline{z}_1 \tag{C.17}$$

holds. This can be reformulated as $0 \leq \underline{z}_1^{\mathsf{T}} \underline{\underline{N}} \underline{z}_1 \leq \underline{z}_1^{\mathsf{T}} \underline{\underline{N}}^{+} \underline{z}_1 \, \forall \underline{z}_1$ with $\underline{\underline{N}} = \underline{\underline{M}} - \underline{\underline{M}}^{-}$ and $\underline{\underline{N}}^{+} = \underline{\underline{M}}^{+} - \underline{\underline{M}}^{-}$. Based on Eq. (C.16), $0 \leq \underline{\underline{N}} \leq \underline{\underline{N}}^{+}$ and $\underline{\underline{M}}^{-} \leq \underline{\underline{M}} \leq \underline{\underline{M}}^{+}$ follow. Bounds for all coefficients of $\underline{\underline{M}}$ can now be obtained with Eq. (C.5).

Appendix D

Stationary HS polarization field

We start at the stationarity condition Eq. (3.88). Reformulation of Eq. (3.88) delivers

$$\boldsymbol{\tau} = \mathbb{R}[\check{\boldsymbol{\varepsilon}} + \mathbb{P}_0[\langle \boldsymbol{\tau} \rangle_v]] \,, \quad \mathbb{R} = (\hat{\mathbb{C}}^{-1} + \mathbb{P}_0)^{-1} \tag{D.1}$$

Averaging, isolating $\langle \boldsymbol{\tau} \rangle_v$ and reinserting $\tilde{\boldsymbol{\varepsilon}} = \bar{\boldsymbol{\varepsilon}} - \hat{\mathbb{C}}^{-1}[\hat{\boldsymbol{s}}]$ delivers

$$\langle \boldsymbol{\tau} \rangle_v = \langle \mathbb{R}[\tilde{\boldsymbol{\varepsilon}}] \rangle_v + \langle \mathbb{R} \rangle_v \mathbb{P}_0[\langle \boldsymbol{\tau} \rangle_v] \tag{D.2}$$

$$\Rightarrow \langle \boldsymbol{\tau} \rangle_v = (\mathbb{I}^S - \langle \mathbb{R} \rangle_v \mathbb{P}_0)^{-1}[\langle \mathbb{R} \rangle_v [\bar{\boldsymbol{\varepsilon}}] - \langle \mathbb{R}\hat{\mathbb{C}}^{-1}[\hat{\boldsymbol{s}}] \rangle_v] \,. \tag{D.3}$$

The tensor $\mathbb{R}$ may be manipulated as

$$\mathbb{R} = (\hat{\mathbb{C}}^{-1}[\mathbb{P}_0^{-1} + \hat{\mathbb{C}}]\mathbb{P}_0)^{-1} = \mathbb{P}_0^{-1}\mathbb{L}\hat{\mathbb{C}} \,, \quad \mathbb{L} = \hat{\mathbb{C}} + \mathbb{P}_0^{-1} \tag{D.4}$$

This identity allows to simplify $\langle \mathbb{R}\hat{\mathbb{C}}^{-1}[\hat{\boldsymbol{s}}] \rangle_v$ as

$$\langle \mathbb{R}\hat{\mathbb{C}}^{-1}[\hat{\boldsymbol{s}}] \rangle_v = \mathbb{P}_0^{-1}[\langle \mathbb{L}[\hat{\boldsymbol{s}}] \rangle_v] \,. \tag{D.5}$$

The tensor $\mathbb{R}$ can be further reformulated as

$$\mathbb{R} = \mathbb{P}_0^{-1}\mathbb{L}\hat{\mathbb{C}} = \mathbb{P}_0^{-1}\mathbb{L}(\mathbb{L}^{-1} - \mathbb{P}_0^{-1}) = \mathbb{P}_0^{-1}(\mathbb{P}_0 - \mathbb{L})\mathbb{P}_0^{-1} \,. \tag{D.6}$$

The average of $\mathbb{R}$ yields

$$\langle \mathbb{R} \rangle_v = \mathbb{P}_0^{-1}(\mathbb{P}_0 - \langle \mathbb{L} \rangle_v)\mathbb{P}_0^{-1} \,, \tag{D.7}$$

such that $\langle \boldsymbol{\tau} \rangle_v$ is computed as

$$\begin{aligned}
\langle \boldsymbol{\tau} \rangle_v &= \langle \mathbb{L} \rangle_v^{-1} \mathbb{P}_0 [\mathbb{P}_0^{-1} (\mathbb{P}_0 - \langle \mathbb{L} \rangle_v) \mathbb{P}_0^{-1} [\bar{\varepsilon}] - \mathbb{P}_0^{-1} [\langle \mathbb{L}[\hat{s}] \rangle_v]] & \text{(D.8)} \\
&= (\langle \mathbb{L} \rangle_v^{-1} - \mathbb{P}_0^{-1}) [\bar{\varepsilon}] - \langle \mathbb{L} \rangle_v^{-1} [\langle \mathbb{L}[\hat{s}] \rangle_v] \,. & \text{(D.9)}
\end{aligned}$$

The local polarization yields

$$\begin{aligned}
\boldsymbol{\tau} &= \mathbb{R}[\bar{\varepsilon} - \hat{\mathbb{C}}^{-1}[\hat{s}] + \mathbb{P}_0 \{ (\langle \mathbb{L} \rangle_v^{-1} - \mathbb{P}_0^{-1}) [\bar{\varepsilon}] \\
&\quad - \langle \mathbb{L} \rangle_v^{-1} [\langle \mathbb{L}[\hat{s}] \rangle_v] \}] & \text{(D.10)} \\
&= \mathbb{R} \mathbb{P}_0 \langle \mathbb{L} \rangle_v^{-1} [\bar{\varepsilon} - \langle \mathbb{L}[\hat{s}] \rangle_v] - \mathbb{R} \hat{\mathbb{C}}^{-1} [\hat{s}] & \text{(D.11)} \\
&= (\mathbb{I}^S - \mathbb{P}_0^{-1} \mathbb{L}) \langle \mathbb{L} \rangle_v^{-1} [\bar{\varepsilon} - \langle \mathbb{L}[\hat{s}] \rangle_v] - \mathbb{P}_0^{-1} \mathbb{L}[\hat{s}] \,, & \text{(D.12)}
\end{aligned}$$

where in the last line (D.6) the identities (D.4) and (D.6) have been used. The determined polarization field yields the stationary value of the HS functional. For its evaluation, the scalar product

$$\langle \boldsymbol{\tau} \cdot \tilde{\varepsilon} \rangle_v = \langle \boldsymbol{\tau} \rangle_v \cdot \bar{\varepsilon} - \langle \boldsymbol{\tau} \cdot \hat{\mathbb{C}}^{-1} [\hat{s}] \rangle_v \qquad \text{(D.13)}$$

needs to be computed. The first term evaluates to

$$\bar{\varepsilon} \cdot \langle \boldsymbol{\tau} \rangle_v = \bar{\varepsilon} \cdot (\langle \mathbb{L} \rangle_v^{-1} - \mathbb{P}_0^{-1}) [\bar{\varepsilon}] - \bar{\varepsilon} \cdot \langle \mathbb{L} \rangle_v^{-1} [\langle \mathbb{L}[\hat{s}] \rangle_v] \,. \qquad \text{(D.14)}$$

The second term may be computed based on the identities

$$\begin{aligned}
\hat{\mathbb{C}}^{-1} = (\mathbb{L}^{-1} - \mathbb{P}_0)^{-1} = \mathbb{L} (\mathbb{I}^S - \mathbb{P}_0^{-1} \mathbb{L})^{-1} & \qquad \text{(D.15)} \\
\hat{\mathbb{C}}^{-1} - \mathbb{L} = \hat{\mathbb{C}}^{-1} \mathbb{P}_0^{-1} \mathbb{L} & \qquad \text{(D.16)}
\end{aligned}$$

such that

$$\begin{aligned}
\langle \hat{s} \cdot \hat{\mathbb{C}}^{-1}[\tau] \rangle_v &= \langle \hat{s} \cdot \hat{\mathbb{C}}^{-1}(\mathbb{I}^S - \mathbb{P}_0^{-1}\mathbb{L})\langle \mathbb{L} \rangle_v^{-1}[\bar{\varepsilon} - \langle \mathbb{L}[\hat{s}] \rangle_v] \rangle_v \\
&\quad - \langle \hat{s} \cdot \hat{\mathbb{C}}^{-1}\mathbb{P}_0^{-1}\mathbb{L}[\hat{s}] \rangle_v \qquad \text{(D.17)} \\
&= \langle s \cdot \mathbb{L}\langle \mathbb{L} \rangle_v^{-1}[\bar{\varepsilon} - \langle \mathbb{L}[\hat{s}] \rangle_v] \rangle_v \\
&\quad - \langle \hat{s} \cdot (\hat{\mathbb{C}}^{-1} - \mathbb{L})[\hat{s}] \rangle_v \qquad \text{(D.18)} \\
&= (\bar{\varepsilon} - \langle \mathbb{L}[\hat{s}] \rangle_v) \cdot \langle \mathbb{L} \rangle_v^{-1}[\langle \mathbb{L}[\hat{s}] \rangle_v] \\
&\quad - \langle \hat{s} \cdot (\hat{\mathbb{C}}^{-1} - \mathbb{L})[\hat{s}] \rangle_v \qquad \text{(D.19)}
\end{aligned}$$

holds. The HS potential evaluates to, see Eq. (3.92),

$$\begin{aligned}
W^{HS} &= \tilde{W}_0 + \frac{1}{2}\langle \tau \cdot \tilde{\varepsilon} \rangle_v \qquad \text{(D.20)} \\
&= \left\{ \frac{1}{2}\bar{\varepsilon} \cdot \mathbb{C}_0[\bar{\varepsilon}] - \bar{\varepsilon} \cdot s_0 - \frac{1}{2}\langle k + \hat{s} \cdot \hat{\mathbb{C}}^{-1}[\hat{s}] \rangle_v \right\} \\
&\quad + \frac{1}{2}\left\{ \bar{\varepsilon} \cdot (\langle \mathbb{L} \rangle_v^{-1} - \mathbb{P}_0^{-1})[\bar{\varepsilon}] \right. \\
&\qquad - (2\bar{\varepsilon} - \langle \mathbb{L}[\hat{s}] \rangle_v) \cdot \langle \mathbb{L} \rangle_v^{-1}[\langle \mathbb{L}[\hat{s}] \rangle_v] \\
&\qquad \left. + \langle \hat{s} \cdot (\hat{\mathbb{C}}^{-1} - \mathbb{L})[\hat{s}] \rangle_v \right\} \qquad \text{(D.21)} \\
&= \frac{1}{2}\bar{\varepsilon} \cdot (\mathbb{C}_0 - \mathbb{P}_0^{-1} + \langle \mathbb{L} \rangle_v^{-1})[\bar{\varepsilon}] \\
&\quad - \bar{\varepsilon} \cdot (s_0 + \langle \mathbb{L} \rangle_v^{-1}[\langle \mathbb{L}[\hat{s}] \rangle_v]) \\
&\quad - \frac{1}{2}\left(\langle k \rangle_v + \langle \hat{s} \cdot \mathbb{L}[\hat{s}] \rangle_v - \langle \mathbb{L}[\hat{s}] \rangle_v \cdot \langle \mathbb{L} \rangle_v^{-1}[\langle \mathbb{L}[\hat{s}] \rangle_v] \right) \qquad \text{(D.22)} \\
&= \frac{1}{2}\bar{\varepsilon} \cdot (\mathbb{C}_0 - \mathbb{P}_0^{-1} + \langle \mathbb{L} \rangle_v^{-1})[\bar{\varepsilon}] - \bar{\varepsilon} \cdot \langle \mathbb{L} \rangle_v^{-1}[\langle \mathbb{L}[s] \rangle_v] \\
&\quad - \frac{1}{2}\left(\langle k \rangle_v + \langle s \cdot \mathbb{L}[\hat{]} \rangle_v - \langle \mathbb{L}[s] \rangle_v \cdot \langle \mathbb{L} \rangle_v^{-1}[\langle \mathbb{L}[s] \rangle_v] \right) \qquad \text{(D.23)}
\end{aligned}$$

Frequently used acronyms, symbols, and operators

Acronyms

n-PPF	n-point probability function
CCODF	Central crystallite orientation distribution function
CODF	Crystallite orientation distribution function
CUB	Cubic
FCC	Face centered cubic
HEX	Hexagonal
HS	Hashin-Shtrikman
ISC	Isotropic self-consistent
ISO	Isotropic
LCC	Linear comparison composite
MPPM	Multiphase polycrystalline material
ORT	Orthotropic
PSD	Positive semidefinite
RVE	Representative volume element
SA1	Singular approximation based on Hill average of first-order bounds for isotropic texture
SC	Self-consistent
SPPM	Single phase polycrystalline material
TFE	Tensorial Fourier expansion
TRI	Triclinic

Latin letters

$x, y, W, \ldots$	Scalar quantities
$a_i, A_{ij}, A_{ijkl}, \ldots$	Tensor components in respect to orthonormal basis $\{\boldsymbol{b}_i\}$
$\boldsymbol{x}, \boldsymbol{y}, \ldots$	Vectors
$\boldsymbol{A}, \boldsymbol{B}, \ldots$	Second-order tensors
$\mathbb{A}, \mathbb{B}, \ldots$	Fourth-order tensors
$\mathbb{A}_{\langle n \rangle}, \mathbb{B}_{\langle n \rangle}, \ldots$	n-th-order tensors
$\mathbb{A}$	Strain localization tensor
$\boldsymbol{a}$	Strain fluctuation localization field (second-order tensor)
$\mathbb{B}$	Stress localization tensor
$\boldsymbol{b}_{ij}, \boldsymbol{b}_{ijkl}$	Abbreviations for $\boldsymbol{b}_i \otimes \boldsymbol{b}_j$ and $\boldsymbol{b}_i \otimes \boldsymbol{b}_j \otimes \boldsymbol{b}_k \otimes \boldsymbol{b}_l$
$\{\boldsymbol{b}_1, \boldsymbol{b}_2, \boldsymbol{b}_3\}$	Orthonormal basis vectors of the three dimensional Euclidean physical space
$\boldsymbol{b}$	Stress fluctuation localization field (second-order tensor)
$\mathbb{C}_0$	Constant comparison stiffness
$\hat{\mathbb{C}}$	Stiffness difference $\mathbb{C} - \mathbb{C}_0$
$\mathbb{C}^{n\pm}$	Upper and lower n-th-order bounds of $\bar{\mathbb{C}}$
$\mathbb{C}$	Stiffness tensor
$\tilde{\mathbb{C}}^{0\pm}$	Alternative zeroth-order bounds of linear elastic properties for large material data bases
$\mathbb{C}^{0\pm}$	Zeroth-order bounds of linear elastic properties
$\bar{\mathbb{C}}$	Effective stiffness
$\tilde{\mathbb{C}}$	Single crystal stiffness tensor
$\underline{\underline{C}}^*$	Supertensor based on W^* in complementary space
$\underline{\underline{\tilde{C}}}^{0\pm}$	Alternative supertensor zeroth-order bounds for large material data bases
$\underline{\underline{C}}^{0\pm}$	Supertensor zeroth-order bounds

$\underline{\underline{C}}$	Supertensor based on W
$D_\alpha(\omega)$	Dirichlet kernel
$\tilde{\mathbb{D}}_{\langle r\rangle}$	Single crystal physical property of r-th-order
$\boldsymbol{e}$	Eigenstrain field (second-order tensor)
F	HS functional in elastic space
$\mathbb{F}'_{\langle\alpha\rangle\beta}(\boldsymbol{Q})$	Ansatz function for CODF of α-th-order corresponding to the reference tensor $\mathbb{H}'_{\langle\alpha\rangle\beta}$, $\mathbb{F}'_{\langle\alpha\rangle\beta}(\boldsymbol{Q}) = \boldsymbol{Q} \star \mathbb{H}'_{\langle\alpha\rangle\beta}$
$f(\boldsymbol{Q})$	Crystallite orientation distribution function
e_n	Relative elastic volume based on n-th-order bounds of linear elastic properties
f	Material parameter connected to the heat capacity for constant strain c_ε, $f = \rho c_\varepsilon/\theta_0$
$\mathbb{G}$	Non-local operator based on Green's function
$g(\hat{\omega})$	Central crystallite orientation distribution function
$\mathbb{H}'_{\langle\alpha\rangle}$	Harmonic tensor of α-th-order, i.e., completely symmetric in respect to arbitrary index permutation and fulfilling $\mathbb{H}_{\langle\alpha\rangle}[\boldsymbol{I}] = \mathbb{O}_{\langle\alpha-2\rangle}$
$\mathbb{H}'_{\langle\alpha\rangle\beta}$	Reference tensor of α-th-order for $\beta = 1, \ldots, n'_\alpha$
$\boldsymbol{I}$	Identity on vectors $\boldsymbol{I} = \sum_{i,j=1}^{3} \delta_{ij} \boldsymbol{b}_{ij}$
$\mathbb{I}$	Identity on second-order tensors $\mathbb{I} = \sum_{i,j,k,l=1}^{3} \delta_{ik}\delta_{jl}\boldsymbol{b}_{ijkl}$
$\mathbb{I}^S$	Identity on symmetric second-order tensors $\mathbb{I}^S = \sum_{i,j,k,l=1}^{3} \frac{1}{2}(\delta_{ik}\delta_{jl} + \delta_{il}\delta_{jk})\boldsymbol{b}_{ijkl}$
I^i	Indicator function of phase i
$\mathbb{J}_{\langle r+\alpha\rangle\gamma}$	Constant isotropic $(r+\alpha)$-th-order tensors in harmonic decomposition of $\tilde{\mathbb{D}}_{\langle r\rangle}$
k	Strain independent part of W
l	Stress independent part of U

$\mathcal{N}, \mathcal{N}_\alpha^{\mathrm{sym}}$	Set of harmonic tensors with Frobenius norm smaller or equal to unity and corresponding special case considering tensors up to tensor order α for symmetry group S^{sym}
$\boldsymbol{n}$	Rotation axis
n'_α	Number of non-vanishing harmonic tensors of α-th-order
$Orth$	Set of orthogonal tensors
$\mathbb{O}_{\langle\alpha\rangle}$	Zero α-th-order tensor
$\mathbb{P}_0$	Polarization tensor
$\mathbb{P}_1$	Identity on isotropic second-order tensors $\mathbb{P}_1 = \frac{1}{3}\boldsymbol{I} \otimes \boldsymbol{I}$
$\mathbb{P}_2$	Identity on symmetric, traceless second-order tensors $\mathbb{P}_2 = \mathbb{I}^S - \mathbb{P}_1$
$\mathbb{P}_3, \mathbb{I}^{AS}$	Identity on skewed / antisymmetric second-order tensors $\mathbb{P}_3 = \mathbb{I}^{AS} = \mathbb{I} - \mathbb{I}^S$
Q_n	Equidistant discretization of intervals based on positive integer n for rotation angle and spherical coordinates of rotation axis for orientations
$\boldsymbol{Q}$	Orientation
$\mathbb{Q}^f_{\langle 2r\rangle}, \mathbb{Q}^g_{\langle 2r\rangle}$	Orientation average $(2r)$-th-order tensor based on CODF f or CCODF g
S_2	Unit sphere in 3-dimensional space
$\mathbb{S}$	Compliance tensor
$S_n^{i_1 \dots i_n}$	n-point probability function
$SO(3)$	Set of orientations
$\underline{\underline{S}}$	Supertensor based on U in complementary space
$\bar{\boldsymbol{s}}$	Effective eigenstress
$\boldsymbol{s}$	Eigenstress field (second-order tensor)
U	Complementary potential function
$U^\pm$	Upper and lower bound of $\bar{U}$
$\bar{U}$	Effective complementary potential function

$\boldsymbol{u}$	Displacement field
$\check{\mathcal{V}}, \check{\mathcal{V}}_{\alpha}^{\text{sym}}$	Set of texture coefficients based on convex combinations of single crystal states and corresponding special case considering tensors up to tensor order α for symmetry group S^{sym}
$\mathcal{V}, \mathcal{V}_{\alpha}^{\text{sym}}$	Set of texture coefficients based on general CODF and corresponding special case considering tensors up to tensor order α for symmetry group S^{sym}
$\mathbb{V}'_{\langle\alpha\rangle\beta}$	Texture coefficient of α-th-order corresponding to the reference tensor $\mathbb{H}'_{\langle\alpha\rangle\beta}$
$v, \partial v$	RVE volume region and surface
W	Elastic potential function
$W^{\pm}$	Upper and lower bound of $\bar{W}$
W^{*}	Legendre-Fenchel transform of W
$\bar{W}$	Effective elastic potential function
$\boldsymbol{x}$	Position vector

Greek letters

β	Thermal stress coefficient
$\delta(\hat{\boldsymbol{Q}}, \boldsymbol{Q})$	Dirac distribution in orientation space with central orientation $\hat{\boldsymbol{Q}}$
δ_{ij}	Kronecker symbol
$\boldsymbol{\varepsilon}, \langle\boldsymbol{\varepsilon}\rangle_v, \bar{\boldsymbol{\varepsilon}}$	Infinitesimal strain, volume average over RVE and effective corresponding measure
ε_{ijk}	Permutation symbol
$\boldsymbol{\epsilon}$	Third-order permutation tensor
$\Gamma_{\alpha\beta}^{n\pm}$	Component bounds of stiffness tensor based on n-th-order bounds
$\gamma_{\alpha\beta}^{n\pm}$	Component bounds of supertensors

θ	Spherical coordinate / absolute temperature
λ	Texture eigenvalue / general tensor eigenvalue
$\mu_\alpha(\omega)$	Ansatz function for CCODF based on Dirichlet kernel $D_\alpha(\omega)$
$\boldsymbol{\xi}, \hat{\boldsymbol{\xi}}$	Arbitrary and exact strain polarization
$\bar{\sigma}_f$	Effective flow stress
$\boldsymbol{\sigma}, \langle\boldsymbol{\sigma}\rangle_v, \bar{\boldsymbol{\sigma}}$	Cauchy stress, volume average over RVE and effective corresponding measure
$\boldsymbol{\tau}, \hat{\boldsymbol{\tau}}$	Arbitrary and exact stress polarization
τ_{kc}	Argument of maximal potential difference in k-th slip system
τ_k	Shear stress of k-th slip system
ϕ	Spherical coordinate
ϕ_{kc}	Comparison potential of k-th slip system in LCC
ϕ_k	Potential of k-th slip system
φ_{kc}	Maximal potential difference in k-th slip system
$\Omega, \partial\Omega$	Material region of a sample and surface
$\hat{\omega}$	Orientation distance
ω	Rotation angle

Operators

$\boldsymbol{A}^{\otimes r}$	Dyadic / tensor power, defined by $\boldsymbol{A}^{\otimes 2} = \boldsymbol{A} \otimes \boldsymbol{A}$ and $\boldsymbol{A}^{\otimes r} = \boldsymbol{A}^{\otimes(r-1)} \otimes \boldsymbol{A}$
$\boldsymbol{A} \otimes \boldsymbol{B}$	Dyadic / tensor product $\boldsymbol{A} \otimes \boldsymbol{B} = \sum_{ijkl=1}^{3} A_{ij} B_{kl} \boldsymbol{b}_i \otimes \boldsymbol{b}_j \otimes \boldsymbol{b}_k \otimes \boldsymbol{b}_l$
$\lVert\boldsymbol{A}\rVert$	Frobenius norm $\lVert\boldsymbol{A}\rVert = \sqrt{\boldsymbol{A} \cdot \boldsymbol{A}}$
$\mathbb{A}[\boldsymbol{B}]$	Linear map $\mathbb{A}[\boldsymbol{B}] = \sum_{i,j,k,l=1}^{3} A_{ijkl} B_{kl} \boldsymbol{b}_{ij}$
$\boldsymbol{A}^{\star r}$	Rayleigh power, defined by $\boldsymbol{A}^{\star r}[\mathbb{B}_{\langle r\rangle}] = \boldsymbol{A} \star \mathbb{B}_{\langle r\rangle}$

$\boldsymbol{A} \star \mathbb{B}_{\langle r \rangle}$	Rayleigh product $\boldsymbol{A} \star \mathbb{B}_{\langle r \rangle} = \sum_{i_1 \dots i_r = 1}^{3} B_{i_1 \dots i_r} (\boldsymbol{A}\boldsymbol{b}_{i_1}) \otimes \cdots \otimes (\boldsymbol{A}\boldsymbol{b}_{i_r})$
$\boldsymbol{A} \cdot \boldsymbol{B}$	Scalar product $\boldsymbol{A} \cdot \boldsymbol{B} = \sum_{ij=1}^{3} A_{ij} B_{ij}$
δ	Variation
$\langle q \rangle_f$	Orientation average of a quantity q over $SO(3)$ based on the CODF $f(\boldsymbol{Q})$
$\langle q \rangle_v$	Volume average of a quantity q over the RVE region v
$\langle q \rangle$	Ensemble average of a quantity q over the sample space of the manufacturing process
$\mathbb{A}^{\dagger}$	Adjoint operator
$\mathrm{div}(\mathbb{A}_{\langle r \rangle})$	Divergence of a tensor field $\mathbb{A}_{\langle r \rangle}$ defined as $\mathrm{div}(\mathbb{A}_{\langle r \rangle}) = \mathrm{grad}(\mathbb{A}_{\langle r \rangle})[\boldsymbol{I}]$
$\mathrm{grad}(q)$	Gradient of a field q
$\mathrm{gs}^{\mathrm{sym}}(\mathbb{A}_{\langle r \rangle})$	Group symetrization of the tensor $\mathbb{A}_{\langle r \rangle}$ in respect to the group indexed by sym, e.g., hex for hexagonal materials

Bibliography

Adams, B. L., Boehler, J., Guidi, M., Onat, E., 1992. Group theory and representation of microstructure and mechanical behavior of polycrystals. J. Mech. Phys. Solids 40 (4), 723–737.

Adams, B. L., Henrie, A., Henrie, B., Lyon, M., Kalidindi, S. R., Garmestani, H., 2001. Microstructure-sensitive design of a compliant beam. J. Mech. Phys. Solids 49 (8), 1639–1663.

Adams, B. L., Kalidindi, S. R., Fullwood, D. T., 2013. Microstructure Sensitive Design for Performance Optimization. Butterworth-Heinemann, Waltham, MA.

Andrews, D. L., Ghoul, W. a., 1999. Eighth rank isotropic tensors and rotational averages. J. Phys. A. Math. Gen. 14 (6), 1281–1290.

Ashby, M. F., Johnson, K., 2013. Materials and Design: The Art and Science of Material Selection in Product Design. Butterworth-Heinemann.

Auffray, N., 2015. On the algebraic structure of isotropic generalized elasticity theories. Math. Mech. Solids 20 (5), 565–581.

Beran, M., 1968. Statistical Continuum Theories. Interscience, New York.

Bertram, A., Böhlke, T., Gaffke, N., Heiligers, B., Offinger, R., 2000. On the Generation of Discrete Isotropic Orientation Distributions for Linear Elastic Cubic Crystals. J. Elast. 58 (3), 233–248.

Bishop, J. F. W., Hill, R., 1951. XLVI. A theory of the plastic distortion of a polycrystalline aggregate under combined stresses. Philos. Mag. Ser. 7 42 (327), 414–427.

Böhlke, T., 2001. Crystallographic Texture Evolution and Elastic Anisotropy: Simulation, Modeling, and Applications. Ph.D. thesis, Otto-von-Guericke-Universität Magdeburg.

Böhlke, T., 2004. The Voigt bound of the stress potential of isotropic viscoplastic FCC polycrystals. Arch. Mech. 56 (6), 425–445.

Böhlke, T., 2005. Application of the maximum entropy method in texture analysis. Comput. Mater. Sci. 32 (3-4), 276–283.

Böhlke, T., 2006. Texture simulation based on tensorial Fourier coefficients. Comput. Struct. 84 (17-18), 1086–1094.

Böhlke, T., Bertram, A., 2001a. Isotropic orientation distributions of cubic crystals. J. Mech. Phys. Solids 49 (11), 2459–2470.

Böhlke, T., Bertram, A., 2001b. The evolution of Hooke's law due to texture development in FCC polycrystals. Int. J. Solids Struct. 38 (52), 9437–9459.

Böhlke, T., Bertram, A., 2003. The Reuss bound of the strain rate potential of viscoplastic fcc polycrystals. Tech. Mech. 23 (2-4), 184–194.

Böhlke, T., Haus, U.-U., Schulze, V., 2006. Crystallographic texture approximation by quadratic programming. Acta Mater. 54 (5), 1359–1368.

Böhlke, T., Lobos, M., 2014. Representation of Hashin-Shtrikman bounds of cubic crystal aggregates in terms of texture coefficients with application in materials design. Acta Mater. 67, 324–334.

Bunge, H. J., 1982. Texture Analysis in Materials Science: Mathematical Methods. Butterworth, London.

Clément, A., Coulomb, P., 1979. Eulerian simulation of deformation textures. Scr. Metall. 13 (9), 899–901.

de Jong, M., Chen, W., Angsten, T., Jain, A., Notestine, R., Gamst, A., Sluiter, M., Krishna Ande, C., van der Zwaag, S., Plata, J. J., Toher, C., Curtarolo, S., Ceder, G., Persson, K. A., Asta, M., 2015. Charting the complete elastic properties of inorganic crystalline compounds. Sci. Data 2, 1–13.

DeBotton, G., Ponte Castañeda, P., 1995. Variational estimates for the creep behaviour of polycrystals. Proc. R. Soc. A 448, 121–142.

Dendievel, R., Bonnet, G., Willis, J. R., 1991. Bounds for the Creep Behaviour of Polycrystalline Materials. Springer New York, New York, NY, pp. 175–192.

Du, W., Man, C.-S., 2017. Material Tensors and Pseudotensors of Weakly-Textured Polycrystals with Orientation Distribution Function Defined on the Orthogonal Group. J. Elast. 127 (2), 197–233.

Fokin, A., 1972. Solution of statistical problems in elasticity theory in the singular approximation. J. Appl. Mech. Tech. Phys. 13 (1), 85–89.

Forte, S., Vianello, M., 1996. Symmetry Classes for Elasticity Tensors. J. Elast. 43 (2), 81–108.

Forte, S., Vianello, M., 1997. Symmetry classes and harmonic decomposition for photoelasticity tensors. Int. J. Eng. Sci. 35 (14), 1317–1326.

Fullwood, D. T., Niezgoda, S. R., Adams, B. L., Kalidindi, S. R., 2010. Microstructure sensitive design for performance optimization. Prog. Mater. Sci. 55 (6), 477–562.

Gaffke, N., Heiligers, B., Offinger, R., 2002. Isotropic discrete orientation distributions on the 3D special orthogonal group. Linear Algebra Appl. 354 (1-3), 119–139.

Ganster, J., Gems, D., 1985. Polycrystalline Simple Average of Mechanical Properties in the General (Triclinic) Case. Phys. Status Solidi 132 (2), 395–407.

Gel'fand, I. M., Minlos, R., Shapiro, Z., 1963. Representations of the Rotation and Lorentz Groups and their Applications. Pergamon Press, Oxford.

Guidi, M., Adams, B. L., Onat, E. T., 1992. Tensorial Representation of the Orientation Distribution Function in Cubic Polycrystals. Textures Microstruct. 19 (3), 147–167.

Hashin, Z., Shtrikman, S., 1962. On some variational principles in anisotropic and nonhomogeneous elasticity. J. Mech. Phys. Solids 10 (4), 335–342.

Helming, K., 1997. A nearly equal distant grid of orientations for quantitative texture analysis. Textures Microstruct. 28, 219–230.

Helming, K., 1998. Texture Approximations By Model Components. Mater. Struct. 5 (1), 3–10.

Hershey, A. V., 1954. The elasticity of an isotropic aggregate of anisotropic cubic crystals. J. Appl. Mech. 21 (3), 236–240.

Hill, R., 1952. The elastic behaviour of a crystalline aggregate. Proc. Phys. Soc. Sect. A 65, 349–354.

Hill, R., 1965. A self-consistent mechanics of composite materials. J. Mech. Phys. Solids 13 (4), 213–222.

Horn, R. A., Johnson, C. R., 1990. Matrix Analysis. Cambridge Univ. Press, Cambridge.

Hutchinson, J., 1976. Bounds and self-consistent estimates for creep of polycrystalline materials. Proc. R. Soc. A 348, 101–127.

Jerphagnon, J., Chemla, D., Bonneville, R., 1978. The description of the physical properties of condensed matter using irreducible tensors. Adv. Phys. 27 (4), 609–650.

Kalidindi, S. R., 2015. Hierarchical Materials Informatics. Butterworth-Heinemann, Waltham, MA.

Kalidindi, S. R., Houskamp, J. R., Lyons, M., Adams, B. L., 2004. Microstructure sensitive design of an orthotropic plate subjected to tensile load. Int. J. Plast. 20 (8-9), 1561–1575.

Kalisch, J., Bertram, A., 2013. Fast Alternatives to Taylor and Sachs Models for Rigid Perfectly Viscoplastic Polycrystals. Tech. Mech. 33 (2), 104–118.

Knezevic, M., Beyerlein, I. J., Brown, D. W., Sisneros, T. A., Tomé, C. N., 2013. A polycrystal plasticity model for predicting mechanical response and texture evolution during strain-path changes: Application to beryllium. Int. J. Plast. 49, 185–198.

Kröner, E., 1958. Berechnung der elastischen Konstanten des Vielkristalls aus den Konstanten des Einkristalls. Zeitschrift für Phys. 151 (4), 504–518.

Kröner, E., 1977. Bounds for effective elastic moduli of disordered materials. J. Mech. Phys. Solids 25 (2), 137–155.

Laws, N., 1973. On the thermostatics of composite materials. J. Mech. Phys. Solids 21 (34), 9–17.

Lebensohn, R., Liu, Y., Ponte Castañeda, P., 2004. On the accuracy of the self-consistent approximation for polycrystals: Comparison with full-field numerical simulations. Acta Mater. 52 (18), 5347–5361.

Li, D., Garmestani, H., Ahzi, S., 2007. Processing path optimization to achieve desired texture in polycrystalline materials. Acta Mater. 55 (2), 647–654.

Liu, Y., Ponte Castañeda, P., 2004. Second-order theory for the effective behavior and field fluctuations in viscoplastic polycrystals. J. Mech. Phys. Solids 52 (2), 467–495.

Lobos, M., Böhlke, T., 2016. On optimal zeroth-order bounds of linear elastic properties of multiphase materials and application in materials design. Int. J. Solids Struct. 84, 40–48.

Lobos, M., Yuzbasioglu, T., Böhlke, T., 2015. Robust materials design of anisotropic elastic properties of polycrystalline composites. Conf. Proc. YIC GACM 2015 , 158–161.

Lobos, M., Yuzbasioglu, T., Böhlke, T., 2017. Homogenization and Materials Design of Anisotropic Multiphase Linear Elastic Materials Using Central Model Functions. J. Elast. 128 (1), 17–60.

Lobos Fernández, M., Böhlke, T., 2018. Representation of Hashin-Shtrikman Bounds in Terms of Texture Coefficients for Arbitrarily Anisotropic Polycrystalline Materials. J. Elast. , Submitted.

Lourenço, B. F., Fukuda, E. H., Fukushima, M., 2016. Optimality conditions for nonlinear semidefinite programming via squared slack variables. Math. Program. , 1–24.

Lücke, K., Pospiech, J., Virnich, K., Jura, J., 1981. On the problem of the reproduction of the true orientation distribution from pole figures. Acta Metall. 29 (1), 167–185.

Man, C.-S., Huang, M., 2011. A Simple Explicit Formula for the Voigt-Reuss-Hill Average of Elastic Polycrystals with Arbitrary Crystal and Texture Symmetries. J. Elast. 105 (1-2), 29–48.

Man, C.-S., Huang, M., 2012. A representation theorem for material tensors of weakly-textured polycrystals and its applications in elasticity. J. Elast. 106 (1), 1–42.

Mardia, K. V., Jupp, P. E., 2008. Directional Statistics. John Wiley & Sons, Inc., London.

Matthies, S., Muller, J., Vinel, G., 1988. On the Normal Distribution in the Orientation Space. Textures Microstruct. 10 (1), 77–96.

Milton, G. W., 2002. The Theory of Composites. Vol. 6. Cambridge University Press, Cambridge.

Morawiec, A., 1989. Calculation of Polycrystal Elastic Constants from Single-Crystal Data. Phys. Status Solidi 154 (2), 535–541.

Morawiec, A., 1994. Review of deterministic methods of calculation of polycrystal elastic constants. Texture Microstruct. 22 (3), 139–167.

Morawiec, A., 1996. The Effective Elastic Constants of Quasi-Isotropic Polycrystalline Materials Composed of Cubic Phases. Phys. Stat. Sol. 155, 353–364.

Morawiec, A., 2004. Orientations and Rotations: Computations in Crystallographic Textures. Springer, Berlin and Heidelberg.

Morawiec, A., Pospiech, J., 1992. Functions Describing Orientation Correlations in Polycrystalline Materials. Textures Microstruct. 19 (1-2), 67–74.

Morawiec, A., Wierzbanowski, K., Jura, J., Baczmański, A., 1991. Prediction of deformation texture in polycrystals. Philos. Mag. A 64 (6), 1251–1263.

Müller, V., Böhlke, T., 2016. Prediction of effective elastic properties of fiber reinforced composites using fiber orientation tensors. Compos. Sci. Technol. 130, 36–45.

Nadeau, J., Ferrari, M., 2001. On optimal zeroth-order bounds with application to Hashin-Shtrikman bounds and anisotropy parameters. Int. J. Solids Struct. 38 (44-45), 7945–7965.

Nebozhyn, M. V., Gilormini, P., Ponte Castañeda, P., 2001. Variational self-consistent estimates for cubic viscoplastic polycrystals: the effects of grain anisotropy and shape. J. Mech. Phys. Solids 49 (2), 313–340.

Nomura, S., Kawai, H., Kimura, I., Kagiyama, M., 1970. General description of orientation factors in terms of expansion of orientation distribution function in a series of spherical harmonics. J. Polym. Sci. Part A-2 Polym. Phys. 8 (3), 383–400.

Ponte Castañeda, P., 1991. The effective mechanical properties of nonlinear isotropic composites. J. Mech. Phys. Solids 39 (1), 45–71.

Ponte Castañeda, P., 1996. Exact Second-Order Estimates for the Effective Mechanical Properties of Nonlinear Composite Materials. J. Mech. Phys. Solids 44 (6), 827–862.

Ponte Castañeda, P., 2002. Second-order homogenization estimates for nonlinear composites incorporating field fluctuations: I-theory. J. Mech. Phys. Solids 50 (4), 737–757.

Ponte Castañeda, P., 2015. Fully optimized second-order variational estimates for the macroscopic response and field statistics in viscoplastic crystalline composites. Proc. R. Soc. A Math. Phys. Eng. Sci. 471, 1–20.

Ponte Castañeda, P., 2016. Stationary variational estimates for the effective response and field fluctuations in nonlinear composites. J. Mech. Phys. Solids 96, 660–682.

Ponte Castañeda, P., Suquet, P., 1997. Nonlinear composites. Adv. Appl. Mech. 34, 171–302.

Ponte Castañeda, P., Telega, J., Gambin, B., 2004. Nonlinear Homogenization and its Applications to Composites, Polycrystals and Smart Materials. Springer, New York.

Proust, G., Kalidindi, S. R., 2006. Procedures for construction of anisotropic elastic-plastic property closures for face-centered cubic polycrystals using first-order bounding relations. J. Mech. Phys. Solids 54 (8), 1744–1762.

Reuss, A., 1929. Berechnung der Fließgrenze von Mischkristallen auf Grund der Plastizitätsbedingung für Einkristalle. ZAMM - J. Appl. Math. Mech. 9 (1), 49–58.

Roe, R.-J., 1965. Description of Crystallite Orientation in Polycrystalline Materials. III. General Solution to Pole Figure Inversion. J. Appl. Phys. 36 (6), 2024.

Rosca, D., Morawiec, A., De Graef, M., 2014. A new method of constructing a grid in the space of 3D rotations and its applications to texture analysis. Model. Simul. Mater. Sci. Eng. 22, 1–17.

Rosen, B. W., Hashin, Z., 1970. Effective thermal expansion coefficients and specific heats of composite materials. Int. J. Eng. Sci. 8, 157–173.

Savyolova, T., 1984. Grain distribution functions with respect to orientations in polycrystals and their Gaussian approximations. Zavod. Lab. 50 (5), 48–52.

Schaeben, H., 1990. Parameterizations and Probability Distributions of Orientations. Textures Microstruct. 13 (1), 51–54.

Schaeben, H., 1992. "Normal" Orientation Distributions. Textures Microstruct. 19 (4), 197–202.

Schaeben, H., 1996. Texture Approximation or Texture Modelling with Components Represented by the von Mises-Fisher Matrix Distribution on SO(3) and the Bingham Distribution on S^4_+. J. Appl. Crystallogr. 29 (5), 516–525.

Schaeben, H., van den Boogaart, K. G., 2003. Spherical harmonics in texture analysis. Tectonophysics 370 (1), 253–268.

Schouten, J. A., 1924. Der Ricci-Kalkül. Springer, Berlin.

Spencer, A., 1970. A note on the decomposition of tensors into traceless symmetric tensors. Int. J. Eng. Sci. 8, 475–481.

Talbot, D. R. S., Willis, J. R., 1985. Variational principles for inhomogeneous non-linear media. IMA J. Appl. Math. 35 (1), 39–54.

Talbot, D. R. S., Willis, J. R., 1987. Bounds and self-consistent estimates for the overall properties of nonlinear composites. IMA J. Appl. Math. (Institute Math. Its Appl. 39 (3), 215–240.

Taylor, G. I., 1938. Plastic strain in metals. J. Inst. Met. 62, 307–324.

Torquato, S., 2002. Random Heterogeneous Materials: Microstructure and Macroscopic Properties. Springer, New York.

Voigt, W., 1910. Lehrbuch der Kristallphysik (mit Ausschluss der Kristalloptik). Teubner Leipzig.

Walpole, L. J., 1966. On bounds for the overall elastic moduli of inhomogeneous systems - I. J. Mech. Phys. Solids 14 (3), 151–162.

Wassermann, G., Grewen, J., 1962. Texturen metallischer Werkstoffe, 2nd Edition. Springer, Berlin and Heidelberg.

Wigner, E. P., 1931. Gruppentheorie und ihre Anwendung auf die Quantenmechanik der Atomspektrum. Vieweg+Teuber Verlag.

Willis, J. R., 1977. Bounds and self-consistent estimates for the overall properties of anisotropic composites. J. Mech. Phys. Solids 25 (3), 185–202.

Willis, J. R., 1981. Variational and Related Methods for the Overall Properties of Composites. Adv. Appl. Mech. 21, 1–78.

Willis, J. R., 1983. The Overall Elastic Response of Composite Materials. J. Appl. Mech. 50 (4b), 1202–1209.

Zheng, Q.-S., Fu, Y.-B., 2001. Orientation Distribution Functions for Microstructures of Heterogeneous m Materials (II) - Crystal Distribution Functions and Irreducible Tensors Restricted by Various Material Symmetries. Appl. Math. Mech. 22 (8), 885–902.

Zheng, Q.-S., Zou, W.-N., 2000. Irreducible decompositions of physical tensors of high orders. J. Eng. Math. 37 (1-3), 273–288.

Zou, W.-N., Zheng, Q.-S., Du, D.-X., Rychlewski, J., 2001. Orthogonal irreducible decompositions of tensors of high order. Math. Mech. Solids 6 (3), 249–267.

Curriculum vitae

Personal data

- Name: Lobos Fernández, Mauricio
- Date of birth: 02.06.1987
- Nationality: Chilean

Chronological educational and academic career:

- 2008-2011: Bachelor of Science in Mechanical Engineering at the Karlsruhe Institute of Technology (KIT), Germany
- 2011-2013: Master of Science in Mechanical Engineering at the Karlsruhe Institute of Technology (KIT), Germany
- 2013-2018: Scientific employee in the Institute for Mechanical Engineering (Prof. Böhlke) at the Karlsruhe Institute of Technology (KIT), Germany
- 01.02.-30.04.2016: Research stay in Cambridge, England, at the Department for Applied Mathematics and Theoretical Physics (DAMTP) visiting Prof. John R. Willis (funded by the Karlsruhe House of Young Scientist (KHYS))

Nominations and awards

- 2012: Research Student Award (RSA), received from the Karlsruhe Institute of Technology (KIT), Germany
- 2014: Fritz-Weidenhammer prize for excellent teaching, received from the Karlsruhe Institute of Technology (KIT), Germany
- 2015: Nomination for Ars Legendi prize for excellent academic teaching in the category "Digital teaching and learning", Germany

Schriftenreihe Kontinuumsmechanik im Maschinenbau
Karlsruher Institut für Technologie (KIT)
(ISSN 2192-693X)

Herausgeber: Prof. Dr.-Ing. Thomas Böhlke

Band 1 Felix Fritzen
Microstructural modeling and computational homogenization of the physically linear and nonlinear constitutive behavior of micro-heterogeneous materials. 2011
ISBN 978-3-86644-699-1

Band 2 Rumena Tsotsova
Texturbasierte Modellierung anisotroper Fließpotentiale. 2012
ISBN 978-3-86644-764-6

Band 3 Johannes Wippler
Micromechanical finite element simulations of crack propagation in silicon nitride. 2012
ISBN 978-3-86644-818-6

Band 4 Katja Jöchen
Homogenization of the linear and non-linear mechanical behavior of polycrystals. 2013
ISBN 978-3-86644-971-8

Band 5 Stephan Wulfinghoff
Numerically Efficient Gradient Crystal Plasticity with a Grain Boundary Yield Criterion and Dislocation-based Work-Hardening. 2014
ISBN 978-3-7315-0245-6

Band 6 Viktor Müller
Micromechanical modeling of short-fiber reinforced composites. 2016
ISBN 978-3-7315-0454-2

Die Bände sind unter www.ksp.kit.edu als PDF frei verfügbar oder als Druckausgabe bestellbar.

Band 7 Florian Rieger
Work-hardening of dual-phase steel. 2016
ISBN 978-3-7315-0513-6

Band 8 Vedran Glavas
Micromechanical Modeling and Simulation of Forming Processes. 2017
ISBN 978-3-7315-0602-7

Band 9 Eric Bayerschen
Single-crystal gradient plasticity with an accumulated plastic slip: Theory and applications. 2017
ISBN 978-3-7315-0606-5

Band 10 Bartholomäus Brylka
Charakterisierung und Modellierung der Steifigkeit von langfaserverstärktem Polypropylen. 2017
ISBN 978-3-7315-0680-5

Band 11 Rudolf Neumann
Two-Scale Thermomechanical Simulation of Hot Stamping. 2017
ISBN 978-3-7315-0714-7

Band 12 Mauricio Lobos Fernández
Homogenization and materials design of mechanical properties of textured materials based on zeroth-, first- and second-order bounds of linear behavior. 2018
ISBN 978-3-7315-0770-3
